WATER RESOURCE PLANNING, DEVELOPMENT AND MANAGEMENT

NATIONAL WATER PROGRAM STRATEGY FOR RESPONDING TO CLIMATE CHANGE

WATER RESOURCE PLANNING, DEVELOPMENT AND MANAGEMENT

Additional books in this series can be found on Nova's website under the Series tab.

Additional E-books in this series can be found on Nova's website under the E-book tab.

CLIMATE CHANGE AND ITS CAUSES, EFFECTS AND PREDICTION

Additional books in this series can be found on Nova's website under the Series tab.

Additional E-books in this series can be found on Nova's website under the E-book tab.

WATER RESOURCE PLANNING, DEVELOPMENT AND MANAGEMENT

NATIONAL WATER PROGRAM STRATEGY FOR RESPONDING TO CLIMATE CHANGE

JUN SONODA
EDITOR

nova publishers
New York

For permission to use material from this book please contact us:
Telephone 631-231-7269; Fax 631-231-8175
Web Site: http://www.novapublishers.com

Additional color graphics may be available in the e-book version of this book.

Library of Congress Cataloging-in-Publication Data

ISBN: 978-1-62618-124-3

Published by Nova Science Publishers, Inc. † New York

CONTENTS

PREFACE

This book provides an overview of the Environmental Protection Agency's (EPA) National Water Program (NWP). Despite the ongoing effects of climate change, the National Water Program continues to achieve its mission to protect and restore our waters to ensure that drinking water is safe; and that aquatic ecosystems sustain fish, plants, and wildlife, as well as economic, recreational, and subsistence activities. Some of the programs and activities already underway throughout the NWP include protecting healthy watersheds and wetlands; managing stormwater with green infrastructure; and improving the efficiency and sustainability of water infrastructure, including promoting energy and water efficiency, reducing pollutants, and protecting drinking water and public health.

Chapter 1 - Climate change poses significant challenges to water resources and the Environmental Protection Agency's (EPA) National Water Program (NWP). The *NWP 2012 Strategy: Response to Climate Change* addresses climate change in the context of our water programs. It emphasizes assessing and managing risk and incorporating adaptation into core programs. Many of the programs and activities already underway throughout the NWP—such as protecting healthy watersheds and wetlands; managing stormwater with green infrastructure; and improving the efficiency and sustainability of water infrastructure, including promoting energy and water efficiency, reducing pollutants, and protecting drinking water and public health—are even more important to do in light of climate change. However, climate change poses such significant challenges to the nation's water resources that more transformative approaches will be necessary. These include critical reflection on programmatic assumptions and development and implementation of plans to address climate change's challenges.

> EPA Vision: Despite the ongoing effects of climate change, the National Water Program will continue to achieve its mission to protect and restore our waters to ensure that drinking water is safe; and that aquatic ecosystems sustain fish, plants, and wildlife, as well as economic, recreational, and subsistence activities.
>
> EPA National Water Program 2012 Strategy: Response to Climate Change

This *2012 Strategy* articulates such an approach. The reader is advised not to interpret the framing of individual strategic actions that use terms such as "encourage" or "consider" to mean that the NWP doesn't recognize the urgency of action. Rather, we recognize that adaptation is itself transformative and requires a collaborative, problem-solving approach, especially in a resource-constrained environment. Further, "adaptive management" doesn't imply a go-slow or a wait-and-see approach; rather, it is an active approach to understand vulnerability, reduce risk, and prepare for consequences while incorporating new science and lessons learned along the way.

Chapter 2 - This is an edited, reformatted and augmented version of the United States Environmental Protection Agency publication Frequently Asked Questions, National Water Program Strategy: Response to Climate Change, dated December 2012.

In: National Water Program Strategy … ISBN: 978-1-62618-124-3
Editor: Jun Sonoda

Chapter 1

NATIONAL WATER PROGRAM 2012 STRATEGY: RESPONSE TO CLIMATE CHANGE*

United States Environmental Protection Agency

DEFINITIONS OF KEY TERMS

From: *America's Climate Choices* (NRC, 2010a-d)

- **Adapt, Adaptation:** Adjustment in natural or human systems to a new or changing environment that exploits beneficial opportunities or moderates negative effects.
- **Adaptive capacity:** The ability of a system to adjust to climate change (including climate variability and extremes) to moderate potential damages, to take advantage of opportunities, or to cope with the consequences.
- **Mitigation:** An intervention to reduce the causes of changes in climate, such as through reducing emissions of greenhouse gases to the atmosphere.

* This is an edited, reformatted and augmented version of the United States Environmental Protection Agency, dated December, 2012.

- **Resilience:** A capability to anticipate, prepare for, respond to, and recover, from significant multi-hazard threats with minimum damage to social well-being, the economy, and the environment.
- **Risk:** A combination of the magnitude of the potential consequence(s) of climate change impact(s) and the likelihood that the consequence(s) will occur.
- **Stationarity:** The idea that natural systems fluctuate within an unchanging envelope of variability.
- **Vulnerability:** The degree to which a system is susceptible to, or unable to cope with, adverse effects of climate change, including climate variability and extremes. Vulnerability is a function of the character, magnitude, and rate of climate variation to which a system is exposed, its sensitivity, and its adaptive capacity.

FOREWORD

For the past forty years, federal, state, tribal, and local governments have worked diligently to identify and address water pollution problems. As a result, our drinking water is safer, our rivers, lakes, and coastal waters are cleaner, and the health of our wetlands and watersheds is improved.

In 2008, the *EPA National Water Program Strategy: Response to Climate Change* described the emerging scientific consensus on the potential impacts of climate change on water resources. Increasingly, impacts are being observed in communities across the nation and are expected to continue, including:

- Increases in water pollution problems due to warmer air and water temperatures and changes in precipitation patterns;
- Impacts on water infrastructure and aquatic systems due to more extreme weather events;
- Changes to the availability of drinking water supplies;
- Waterbody boundary movement and displacement;
- Changing aquatic biology;
- Collective impacts on coastal areas; and
- Indirect impacts due to unintended consequences of human response to climate change.

Despite increasing understanding of climate change, there still remain questions about the scope and timing of climate change impacts, especially at the local scale where most water-related decisions are made. These challenges require us all to come together to find the tools needed to understand and manage risks and to build resilience of both the built and natural environments.

This ***National Water Program 2012 Strategy: Response to Climate Change*** builds on the momentum gained while implementing the *2008 Strategy*. It provides a road map for where we need to go over the long term and articulates a set of mid-term building blocks, i.e., strategic actions that need to be taken to be a "climate ready" national water program. This *2012 Strategy* emphasizes working collaboratively, developing tools, managing risk, and incorporating adaptation into core programs. Many programs and activities already underway become even more important in light of climate change – including strengthening preparedness for extreme weather events, protecting healthy watersheds and wetlands, managing stormwater with green infrastructure, and improving the sustainability of water infrastructure through energy and water efficiency.

The wider context of climate change-related activity that is underway throughout the nation provides an opportunity to work with partners and stakeholders to achieve the goals of the EPA National Water Program while contributing to broader national goals to sustain the natural resources that support our vibrant economy and our quality of life for current and future generations.

Nancy Stoner
Acting Assistant Administrator for Water

I. Executive Summary

Climate change poses significant challenges to water resources and the Environmental Protection Agency's (EPA) National Water Program (NWP). The *NWP 2012 Strategy: Response to Climate Change* addresses climate change in the context of our water programs. It emphasizes assessing and managing risk and incorporating adaptation into core programs. Many of the programs and activities already underway throughout the NWP—such as protecting healthy watersheds and wetlands; managing stormwater with green infrastructure; and improving the efficiency and sustainability of water infrastructure, including promoting energy and water efficiency, reducing pollutants, and protecting drinking water and public health—are even more important to do in light of climate change. However, climate change poses such significant challenges to the nation's water resources that more transformative approaches will be necessary. These include critical reflection on programmatic assumptions and development and implementation of plans to address climate change's challenges.

> **EPA Vision: Despite the ongoing effects of climate change, the National Water Program will continue to achieve its mission to protect and restore our waters to ensure that drinking water is safe; and that aquatic ecosystems sustain fish, plants, and wildlife, as well as economic, recreational, and subsistence activities.**
>
> EPA National Water Program 2012 Strategy: Response to Climate Change

This *2012 Strategy* articulates such an approach. The reader is advised not to interpret the framing of individual strategic actions that use terms such as "encourage" or "consider" to mean that the NWP doesn't recognize the urgency of action. Rather, we recognize that adaptation is itself transformative and requires a collaborative, problem-solving approach, especially in a resource-constrained environment. Further, "adaptive management" doesn't imply a go-slow or a wait-and-see approach; rather, it is an active approach to understand vulnerability, reduce risk, and prepare for consequences while incorporating new science and lessons learned along the way.

A. The Evolving Context

The first *National Water Program Strategy: Response to Climate Change* was published in 2008; it identified 44 key actions that could be taken in the near term to begin to understand and address the potential impacts of climate change on water resources and EPA's mission. This *2012 Strategy* builds on the momentum gained since then; it describes a set of long-term goals for the management of sustainable water resources for future generations in light of climate change, and charts the key "building blocks," (i.e., strategic actions) that would need to be taken to achieve those goals. It also reflects the wider context of climate change-related activity that is underway throughout the nation. The *2012 Strategy* is intended to be a roadmap to guide future programmatic planning and inform decision-makers during the Agency's annual planning process. It describes an array of important actions that should be taken to be a "climate ready" national water program.

IMPACTS OF CLIMATE CHANGE ON WATER RESOURCES

- **Increases in water pollution problems due to warmer air and water temperatures and changes in precipitation patterns**, causing an increase in the number of waters categorized as "impaired," with associated impacts on human health and aquatic ecosystems.
- **Impacts on water infrastructure and aquatic systems due to more extreme weather events**, including heavier precipitation and tropical and inland storms.
- **Changes in the availability of drinking water supplies** due to increased frequency, severity and duration of drought, changing patterns of precipitation and snowmelt, increased evaporation, and aquifer saltwater intrusion, affecting public water supply, agriculture, industry, and energy production uses.
- **Water body boundary movement and displacements** as rising sea levels alter ocean and estuarine shorelines and as changes in water flow, precipitation, and evaporation affect the size of wetlands and lakes.
- **Changing aquatic biology** due to warmer water and changing flows, resulting in deterioration of aquatic ecosystem health in some areas.
- **Collective impacts on coastal areas** resulting from a combination of sea level rise, increased damage from floods and storms, coastal erosion, saltwater intrusion to drinking water supplies, and increasing temperature and acidification of the oceans.
- **Indirect impacts** due to unintended consequences of human response to climate change, such as those resulting from, for example, armoring shorelines or carbon sequestration and other greenhouse gas reduction strategies.

A cross-Agency workgroup embraced 10 guiding principles to inform development of the revised and updated *NWP 2012 Strategy*. In addition, the *2012 Strategy* is designed to reflect the findings of the Interagency Climate Change Adaptation Task Force (ICCATF) and includes EPA's commitments under three climate change strategic plans under development within the federal government for:

- Freshwater resources by the ICCATF Freshwater Work Group.
- The ocean, coasts, and Great Lakes by the National Ocean Council (NOC).
- Fish, wildlife, and plants by the National Fish, Wildlife and Plants Climate Adaptation Workgroup.

This *2012 Strategy* is also intended to be consistent with EPA's broader adaptation planning. Recognizing that climate change is one stressor among many others that water resource managers are grappling with, this strategy is also designed to build on other initiatives, such as the recent *Coming Together for Clean Water* strategy and EPA's *Clean Water and Safe Drinking Water Infrastructure Sustainability Policy*.

B. Programmatic Visions, Goals, and Strategic Actions

The core programmatic elements of this strategy include:

- Infrastructure
- Watersheds and Wetlands
- Coastal and Ocean Waters
- Protecting Water Quality
- Working with Tribes

Each section addressing these core elements is organized using a three-tier framework: *Vision, Goals, and Strategic Actions*. Each section includes a long-term **Vision**, or outcome, for which EPA may be only one of many actors.

For each Vision, we identify a set of **Goals** that reflects the same long-term timeframe as the Vision. The Goals, however, articulate EPA's mission and role in achieving the Vision, and describe *what* we are trying to achieve.

Table ES-1. Summary of Visions, Goals, and Strategic Actions

Infrastructure: In the face of a changing climate, resilient and adaptable drinking water, wastewater, and stormwater utilities (i.e., the water utility sector) ensure clean and safe water to protect the nation's public health and environment by making smart investment decisions to improve the sustainability of their infrastructure and operations and the communities they serve, while reducing greenhouse gas emissions through greater energy efficiency.	
Goal 1: Build the body of information and tools needed to incorporate climate change into planning and decision making.	**SA1:** Improve access to vetted climate and hydrological science, modeling, and assessment tools through the Climate Ready Water Utilities program.
	SA2: Assist wastewater and water utilities to reduce greenhouse gas emissions and increase long-term sustainability with a combination of energy efficiency, co-generation, and increased use of renewable energy resources.
	SA3: Work with the states and public water systems, particularly small water systems, to identify and plan for climate change challenges to drinking water safety and to assist in meeting health based drinking water standards.
	SA4: Promote sustainable design approaches to provide for the long-term sustainability of infrastructure and operations.
Goal 2: Support Integrated Water Resources Management to sustainably manage water resources.	**SA5:** Understand and promote through technical assistance the use of water supply management strategies.
	SA6: Evaluate and provide technical assistance on the use of water demand management strategies.
	SA7: Increase cross-sector knowledge of water supply climate challenges and develop watershed specific information to inform decision making.
Watersheds & Wetlands: Watersheds are protected, maintained, and restored to provide climate resilience and to preserve the ecological, social, and economic benefits they provide; and the nation's wetlands are maintained and improved using integrated approaches that recognize their inherent value as well as their role in reducing the impacts of climate change.	
Goal 3: Identify, protect, and maintain a network of healthy watersheds and supportive habitat corridor networks.	**SA8:** Develop a national framework and support efforts to protect remaining healthy watersheds and aquatic ecosystems.
	SA9: Collaborate with partners on terrestrial ecosystems and hydrology so that effects on water quality and aquatic ecosystems are considered.
	SA10: Integrate protection of healthy watersheds throughout the NWP core programs.
	SA11: Increase public awareness of the role and importance of healthy watersheds in reducing the impacts of climate change.
Goal 4: Incorporate climate resilience into watershed restoration and floodplain management.	**SA12:** Consider a means of accounting for climate change in EPA funded and other watershed restoration projects.
	SA13: Work with federal, state, interstate, tribal, and local partners to protect and restore the natural resources and functions of riverine and coastal floodplains as a means of building resiliency and protecting water quality.
Goal 5: Watershed protection practices	**SA14:** Encourage states to update their source water delineations, assessments or protection plans to address anticipated climate change impacts.

Table ES-1. (Continued)

incorporate Source Water Protection to protect drinking water supplies.	**SA15:** Continue to support collaborative efforts to increase state and local awareness of source water protection needs and opportunities, and encourage inclusion of source water protection areas in local climate change adaptation initiatives.
Goal 6: EPA incorporates climate change considerations into its wetlands programs, including the Clean Water Act 404 program, as appropriate.	**SA16:** Consider the effects of climate change, as appropriate, when making significant degradation determinations in the CWA Section 404 wetlands permitting and enforcement program.
	SA17: Evaluate, in conjunction with the U.S. Army Corps of Engineers, how wetland and stream compensation projects could be selected, designed, and sited to aid in reducing the effects of climate change.
Goal 7: Improve baseline information on wetland extent, condition, and	**SA18:** Expand wetland mapping by supporting wetland mapping coalitions and training on use of the new federal Wetland Mapping Standard.
	SA19: Produce a statistically valid ecological condition assessment of the nation's wetlands.
performance to inform long term planning and priority setting that takes into account the potential added benefits for climate change adaptation and carbon sequestration.	**SA20:** Work with partners and stakeholders to develop information and tools to support long term planning and priority setting for wetland restoration projects.
Coastal and Ocean Waters: Adverse effects of climate change along with collective stressors and unintended adverse consequences of responses to climate change have been successfully prevented or reduced in the ocean and coastal environment. Federal, tribal, state and local agencies, organizations, and institutions are working cooperatively; and information necessary to integrate climate change considerations into ocean and coastal management is produced, readily available, and used.	
Goal 8: Collaborate so that information and methodologies for ocean and coastal areas are collected, produced, analyzed, and easily available.	**SA21:** Collaborate so that synergy occurs, lessons learned are transferred, federal efforts effectively help local communities, and efforts are not duplicative or at cross-purposes.
	SA22: Work within EPA and with the U.S. Global Change Research Program and other federal, tribal, and state agencies to collect, produce, analyze, and format knowledge and information needed to protect ocean and coastal areas and make it easily available.

Goal 9: Support and build networks of local, tribal, state, regional and federal collaborators to take effective adaptation measures for coastal and ocean environments through EPA's geographically targeted programs.	**SA23:** Work with the NWP's larger geographic programs to incorporate climate change considerations, focusing on both the natural and built environments.
	SA24: Address climate change adaptation and build stakeholder capacity when implementing National Estuary Program Comprehensive Conservation and Management Plans and through the Climate Ready Estuaries Program.
	SA25: Conduct outreach and education, and provide technical assistance to state and local watershed organizations and communities to build adaptive capacity in coastal areas outside the NEP and Large Aquatic Ecosystem programs.
Goal 10: Address climate driven environmental changes in coastal areas and provide that mitigation and adaptation are conducted in an environmentally responsible manner.	**SA26:** Support coastal wastewater, stormwater, and drinking water infrastructure owners and operators in reducing climate risks and encourage adaptation in coastal areas.
	SA27: Support climate readiness of coastal communities, including hazard mitigation, pre-disaster planning, preparedness, and recovery efforts.
	SA28: Support preparation and response planning for impacts to coastal aquatic environments.
Goal 11: Protect ocean environments by incorporating shifting environmental conditions and other emerging threats into EPA programs.	**SA29:** Consider climate change impacts on marine water quality in NWP ocean management authorities, policies, and programs.
	SA30: Use available authorities and work with the regional ocean organizations and other federal and state agencies through regional ocean groups and other networks so that offshore renewable energy production does not adversely affect the marine environment.
	SA31: Support the evaluation of sub-seabed sequestration of carbon dioxide and any proposals for ocean fertilization.
	SA32: Participate in interagency development and implementation of federal strategies through the National Ocean Council and the National Ocean Council Strategic Action Plans.
Water Quality: Our Nation's surface water, drinking water, and ground water quality are protected, and the risks of climate change to human health and the environment are diminished, through a variety of adaptation and mitigation strategies.	
Goal 12: Protect waters of the United States and promote management of sustainable surface water resources.	**SA33:** Encourage states and communities to incorporate climate change considerations into their water quality planning.
	SA34: Encourage green infrastructure and low-impact development to protect water quality and make watersheds more resilient.
	SA35: Promote consideration of climate change impacts by National Pollutant Discharge Elimination System (NPDES) permitting authorities.

Table ES-1. (Continued)

	SA36: Encourage water quality authorities to consider climate change impacts when developing wasteload and load allocations in Total Maximum Daily Loads where appropriate.
	SA37: Identify and protect designated uses that are at risk from climate change impacts.
	SA38: Clarify how to re-evaluate aquatic life water quality criteria on more regular intervals; and develop information to assist states and tribes who are developing criteria that incorporate climate change considerations for hydrologic condition.
Goal 13: As the nation makes decisions to reduce greenhouse gases and develop alternative sources of energy and fuel, work to protect water resources from unintended adverse consequences.	**SA39:** Continue to provide perspective on the water resource implications of new energy technologies.
	SA40: Provide assistance to states and permittees to assure that geologic sequestration of CO_2 is responsibly managed.
	SA41: Continue to work with States to help them identify polluted waters, including those affected by biofuels production, and help them develop and implement Total Maximum Daily Loads (TMDLs) for those waters.
	SA42: Provide informational materials for stakeholders to encourage the consideration of alternative sources of energy and fuels that are water efficient and maintain water quality.
	SA43: As climate change affects the operation or placement of reservoirs, work with other federal agencies and EPA programs to understand the combined effects of climate change and hydropower on flows, water temperature, and water quality.
Goal 14: Collaborate to make hydrological and climate data and projections available.	**SA44:** Monitor climate change impacts to surface waters and ground water.
	SA45: Collaborate with other federal agencies to develop new methods for use of updated precipitation, storm frequency, and observational streamflow data, as well as methods for evaluating projected changes in low flow conditions.
	SA46: Enhance flow estimation using National Hydrography Dataset Plus (NHDPlus).
Working With Tribes: Tribes are able to preserve, adapt, and maintain the viability of their culture, traditions, natural resources, and economies in the face of a changing climate.	
Goal 15: Incorporate climate change considerations in the implementation of core programs, and collaborate with other EPA offices and federal agencies to work with tribes on climate change issues on a multimedia basis.	**SA47:** Through formal consultation and other mechanisms, incorporate climate change as a key consideration in the revised NWP Tribal Strategy and subsequent implementation of Clean Water Act, Safe Drinking Water Act, and other core programs.
	SA48: Incorporate adaptation into tribal funding mechanisms, and collaborate with other EPA and federal funding programs to support sustainability and adaptation in tribal communities.

Goal 16: Tribes have access to information on climate change for decision making.	**SA49:** Collaborate to explore and develop climate change science, information, and tools for tribes, and incorporate local knowledge.
	SA50: Collaborate to develop communication materials relevant for tribal uses and tribal audiences.
Cross-Cutting Program Support	
Goal 17: Communicate, Collaborate, and Train.	**SA51:** Continue building the communication, collaboration, and training mechanisms needed to effectively increase adaptive capacity at the federal, tribal, state, and local levels.
Goal 18: Track Progress and Measure Outcomes	**SA52:** Adopt a phased approach to track programmatic progress towards Strategic Actions; achieve commitments reflected in the Agency's *Strategic Plan*; work with an EPA workgroup to develop outcome measures.
Goal 19: Identify Climate Change and Water Research Needs	**SA53:** Work with EPA's Of fice of Research and Development, other water science agencies, and the water research community to further define needs and develop research opportunities to deliver the information needed to support implementation of this *2012 Strategy*, including providing the decision support tools needed by water resource managers.

Finally, each Goal contains several **Strategic Actions**. The Strategic Actions are the programmatic building blocks to achieve the Goals. These describe *how* the NWP intends to work over the next three to eight years in pursuit of our longer term Goals and Visions.

Table ES-1 summarizes the Visions, Goals, and Strategic Actions described in this *2012 Strategy*. In total, we describe 5 Visions, 19 Goals, and 53 Strategic Actions.

C. Geographic Climate Regions

This section describes the collective strategic focus of EPA Regions working together, organized by the climate impact regions delineated by the U.S. Global Change Research Program (USGCRP), with the addition of a "Montane" region (Table ES-2). Several EPA Regions span multiple USGCRP regions and therefore, each EPA Region will address a variety of climate impacts in its program implementation.

Table ES-2. USGCRP Climate Regions and EPA Regions

Climate Regions	EPA Regions
Northeast	1, 2, 3
Southeast	3, 4, 6
Midwest	2, 5,7
Great Plains	6, 7, 8
Southwest	6, 8, 9
Pacific Northwest	8, 10
Montane	8, 9, 10
Alaska	10
Caribbean Islands	2

TEN GUIDING PRINCIPLES

- **Integrated Water Resources Management (IWRM)**
- **Adaptive Management**
- **Collaborative Learning and Capacity Development**
- **Long Term Planning (i.e., multi-decadal time horizon)**

- **Energy-Water Nexus**
- **Systems & Portfolio Approach**
- **Cost of Inaction**
- **Environmental Justice**
- **Performance Evaluation**
- **Mainstreaming Climate Change into Core Programs**

Successfully achieving the long-term goals will result from strong partnerships with federal agencies, states, interstates, tribes, local governments, nongovernmental, and private sector stakeholders. Specific partnerships in each climate region will vary according to the needs and issues of that region. Of particular importance are the federal efforts underway by the ICCATF to develop "regional consortia" of federal agencies to coordinate delivery of climate services to regional and local stakeholders that include, among others, Landscape Conservation Cooperatives (LCCs) and Climate Science Centers (CSCs) launched by the Department of the Interior, and the National Oceanic and Atmospheric Administration's (NOAA's) Regional Integrated Sciences and Assessments (RISAs) and National Climatic Data Centers.

D. Cross-Cutting Program Support

This section describes essential processes to support and effectively implement the Visions, Goals, and Strategic Actions.

Communication, Collaboration, and Training: The NWP intends to strengthen and expand collaboration, outreach, and training with key partners throughout EPA and with other federal agencies; state, interstate, tribal, and local water program managers; and nongovernmental and private sector stakeholders, using both formal and informal stakeholder involvement opportunities.

Tracking Progress and Measuring Outcomes: Measuring progress toward adaptation is complicated. The current state of practice leans largely to tracking institutional *progress* in incorporating climate change considerations into programs. Similarly, the NWP is developing an approach that evaluates the collectivity of outputs and actions to demonstrate progress in each of several phases toward achieving resilience to climate change, noted in Table ES-3. The NWP intends to work with the State-Tribal Climate Change Council and other partners to refine this approach. As EPA and the ICCATF develop methods for measuring *outcomes*, the NWP intends to incorporate those measures into its evaluation process. In addition, this *2012 Strategy* reflects the NWP's intent to meet the Agency-wide strategic measures adopted in the *EPA 2011–2015 Strategic Plan* and achieve measures embodied in future EPA Strategic Plans.

Climate Change and Water Research Needs: The *2012 Strategy* identifies the types of research needed to support the goals and strategic actions. The NWP intends to continue to work with the EPA's Office of Research and Development (ORD), other water science agencies, and the water research community to further define needs and develop collaborative and coordinated research opportunities.

Conclusion

Climate change alters the hydrological background in which EPA's programs function. In response, EPA intends to evaluate the need to revise data collection, analytical methods, and even regulatory practices that have been developed over the past 40 years since passage of the Clean Water Act (CWA) and the Safe Drinking Water Act (SDWA). This is no easy task; ensuring that EPA's programs continue to protect public health and the environment and sustain the economy calls for immediate and sustained collaboration at the federal, state, interstate, tribal, and local levels.

Table ES-3. Tracking Progress: Phases of Organizational Adaptation

1. Initiation	1
2. Assessment	2
3. Response Development	3
4. Initial Implementation	4
5. Robust Implementation	5
6. Mainstreaming	6
7. Monitoring and Adaptive Management	7

II. Introduction: The Evolving Context

A. *2008 Strategy* vs. *2012 Strategy*

The *National Water Program Strategy: Response to Climate Change*, published in 2008 (*2008 Strategy*), describes the likely effects that climate change will have on water resources and their implications for the EPA's NWP.[1] The *2008 Strategy* laid out 44 "key actions" that the NWP intended to take during 2008–2009, and an update extended the period of action to 2010–2011 (EPA, 2008a).

The *2012 Strategy* builds on the momentum achieved through the implementation of the 44 key actions in the *2008 Strategy*. Further, this *2012 Strategy* describes a longer term vision for the management of sustainable water resources in light of climate change and identifies the key "building blocks" or strategic actions that need to be taken to achieve the long-term goals. It also reflects the wider context of climate change-related activity that is underway throughout the nation. This *2012 Strategy* is a roadmap that reflects directional intention. While it describes an array of important actions consistent with creating a "climate-ready" national water program, it does not outline commitments to act within a specific timeframe. All proposed activities are contingent upon availability of resources and subject to change as new information develops to inform adaptive responses.

B. Relationship of the *2012 Strategy* to Other Planning Activities

The **Interagency Climate Change Adaptation Task Force** (the Task Force) was established under Executive Order 13514 (CEQ, 2009) to develop recommendations for climate change adaptation. On October 5, 2010, the Task Force delivered its initial report to the President with a first set of recommendations (CEQ, 2010a).

Two recommendations in the October 2010 Task Force Report inform the development of the *2012 Strategy*. First, the Task Force's **Freshwater Workgroup**[2] was asked to develop a National Action Plan (NAP) in coordination with similar action plans under development; one by ICCATF's **Fish, Wildlife and Plants Climate Adaptation Workgroup** (FWP Workgroup) and the other by the **National Ocean Council** (NOC) addressing ocean, coastal, and Great Lakes resources. An ICCATF coordinating team has worked together to ensure that the three national adaptation strategies produced by these three workgroups are complementary.

Subsequently, the Freshwater Workgroup published the National Action Plan titled *Priorities for Managing Freshwater Resources in a Changing Climate*[3] (CEQ, 2011a), which describes a National Goal, supported by six recommendations, described below:

INTERAGENCY CLIMATE CHANGE ADAPTATION TASK FORCE

Freshwater National Action Plan

National Goal: Government agencies and citizens work collaboratively to manage freshwater resources in response to a changing climate in order to assure adequate water supplies, to protect human life, health and property, and to protect water quality and aquatic ecosystems.

- **Recommendation # 1:** Establish a Planning Process and Organizational Framework
- **Recommendation # 2:** Improve Water Resources and Climate Change Information
- **Recommendation # 1:** Strengthen Assessment of Vulnerability
- **Recommendation # 4:** Expand Water Use Efficiency
- **Recommendation # 5:** Support Integrated Water Resources Management
- **Recommendation # 6:** Support Training and Outreach to Build Response Capability

—ICCATF Freshwater National Action Plan (CEQ, 2011a)

The Freshwater NAP lays out 24 key actions that support the six recommendations. For some of the supporting actions, EPA will provide leadership, and for those led by other federal agencies, EPA will participate as a team member, as appropriate.

EPA water program staff and managers also participate on the NOC (NOC, 2011) and the Fish, Wildlife, and Plants (FWP) Workgroup (FWP, 2011), and EPA's NWP commitments in those adaptation plans are also reflected in this *2012 Strategy*.

The second recommendation of the Interagency Task Force report called on every federal agency to develop and implement a climate adaptation plan addressing the challenges posed to our missions and operations. The White House Council on Environmental Quality (CEQ) issued implementation instructions on climate adaptation planning to all federal agencies (CEQ, 2011a and b); initial plans were to be submitted by June 2012, and more complete plans submitted by June 2013. In response, EPA established a Policy on Climate Change Adaptation, issued June 2, 2011 (EPA, 2011a), and formed a cross-EPA Work Group on Climate Change Adaptation Planning (EPA Work Group). The Office of Water (OW) and the 10 EPA Regions participate on the EPA Work Group, ensuring that the two Strategies (NWP's and EPA's) are consistent and mutually reinforcing. EPA submitted its plan to CEQ on June 28, 2012.

Despite many successes over recent years, the rate at which waters are being listed for impairment exceeds the rate at which they are being restored. The causes of degradation are in many cases far more complex, and not as visible to the naked eye as they were years ago; the solutions are often available technically, but because the pollution comes from multiple sources, and involves a greater array of pollutants and stressors, it requires new and innovative partnerships and approaches. In some cases EPA and state authorities are limited in scope, and as a result it is challenging to directly address root causes—i.e., population growth, urbanization, agriculture, and other nonpoint source pollution. Building strong and effective partnerships with the widest range of stakeholders, state, local, and tribal partners, and other federal agencies has never been so urgent if we are to protect ourwater and its multiple uses for generations to come.

—Coming Together for Clean Water, (EPA, 2011b)

Additionally, EPA has adopted Agency-wide goals that call for each program office to incorporate climate change science trend and scenario information into five major scientific models and/or decision-support tools; five rulemaking processes; and five major grant, loan, contract, or technical assistance programs, and sets a target for reducing greenhouse gas emissions through energy and resource conservation (EPA, 2010a). This *2012 Strategy* reflects the NWP's commitment to achieving each of these measures by 2015. (See the section on *Tracking Progress and Measuring Outcomes* in Chapter VI for more discussion.)

Finally, in 2010, EPA convened a forum to discuss how to accelerate progress in protecting the nation's waters. The resulting white paper, titled *Coming Together for Clean Water* (CT4CW), recognizes that climate change is just one of the several stressors to water resources (EPA, 2011b). The *Coming Together* strategy presents a framework for how EPA's NWP will work to address today's clean water challenges, such as stormwater, nutrients, and protecting and restoring watersheds. The *Infrastructure Sustainability Policy* reflects EPA's goal to ensure that federal investments, policies, and actions support water infrastructure in efficient and sustainable locations to best aid existing communities, enhance economic competitiveness, and promote affordable neighborhoods. The NWP *2012 Strategy* should be viewed as an indepth treatment of climate change, addressing one of the new and challenging issues facing our program, and as an integral and complementary part of overall NWP strategic planning and initiatives.

C. Impacts of Climate Change on Water Resources: Recent Literature

Recently published assessments and other reports reinforce the findings in the *2008 Strategy* that climate change has significant implications for water resources and water programs. They support EPA's determination that these implications should be addressed in each part of the NWP in order to achieve EPA's mission of protecting human health and the environment. It is important to note that not all impacts of climate change will necessarily be disruptive to particular programmatic endpoints, and that some could at least in the near term provide beneficial opportunities. However, on balance, the range of challenges posed by the interface between built and natural systems and the changing hydrometeorological background conditions is likely to require response actions in order to minimize detrimental effects to current built and natural systems. The impacts listed here refer to the general risks to water resources posed by climate change, but whether and to what degree these risks are likely to be realized in specific locations will require local assessment. The reader is referred to the original *2008 Strategy*, as well as more recent literature cited below and the references cited in Appendix D, for a more detailed discussion of the implications of climate change for water resources and EPA's water programs. These implications include:

- **Increases in water pollution problems:** Warmer air temperatures will result in warmer water. Warmer waters will hold less dissolved oxygen, making instances of low oxygen levels and "hypoxia" (i.e., when dissolved oxygen declines to the point where aquatic species can no longer survive) more likely; foster harmful algal blooms; and change the toxicity of some pollutants (Figure 1).
  The number of waters categorized as "impaired" is likely to increase, even if pollution levels are stable, with associated impacts on human health from waterborne disease and degradation of aquatic ecosystems.
- **Impacts on water infrastructure and aquatic systems due to more extreme weather events (Figure 2):** Heavier precipitation from tropical and inland storms will increase flood risk, expand flood hazard areas, increase the variability of streamflows (i.e., higher high-flows and lower low-flows), increase the velocity of water during high-flow periods, and increase erosion. These changes will have adverse effects on water quality and aquatic ecosystem health. For example, increases in intense rainfall result in more nutrients, pathogens, and toxins being washed into water bodies.
- **Changes to water availability:** In some parts of the country, droughts, changing patterns of precipitation and snowmelt, and increased water loss due to evaporation as a result of warmer temperatures will result in changes to the availability of water for drinking and for use in agriculture, industry, and energy production. In other areas, sea level rise and saltwater intrusion will have the same effect. Warmer air temperatures may also result in increased demands on community water supplies, and the water needs for agriculture, industry, and energy production are likely to increase.

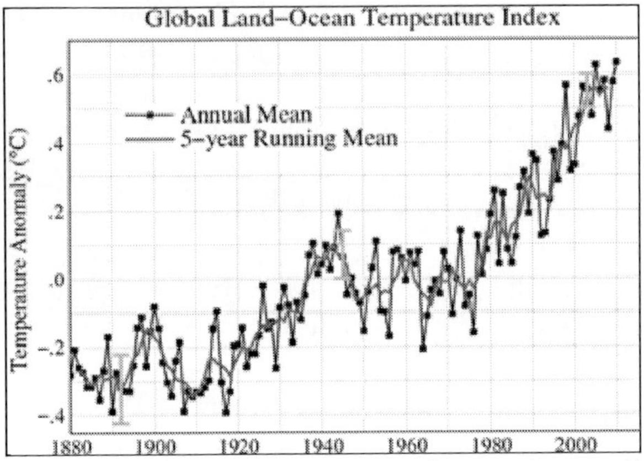

Black curve shows annual average temperatures; red curve shows a five-year running average; green bars indicate the estimated uncertainty in the data during different periods of the record. Source: NASA GISS updated through 2010 at http://data.giss.nasa.gov/gistemp/graphs/.

Figure 1. Global Surface Temperature Change—1880 to 2010 (degrees Celsius)—Compared to Base Period 1951 to 1980.

- **Waterbody boundary movement and displacement:** Rising sea levels will move ocean and estuarine shorelines byinundating lowlands, displacing wetlands, and altering the tidal range in rivers and bays. Changing water flow to lakes and streams, increased evaporation, and changed precipitation in some areas will affect the size of wetlands and lakes. Water levels in the Great Lakes are expected to fall.
- **Changing aquatic biology:** As waters become warmer, the aquatic life they now support will be replaced by other species better adapted to the warmer water (i.e., coldwater fish will be replaced by warmwater fish). This process, however, will occur at an uneven pace, disrupting aquatic system health and allowing nonindigenous and/or invasive species to become established. In the long term (i.e., 50 years), warmer water and changing flows may result in significant deterioration of aquatic ecosystem health in some areas.
- **Collective impacts on coastal areas:** Most areas of the United States will see several water-related impacts, but coastal areas are likely to see multiple impacts associated with climate change (e.g., sea level rise, increased damage from floods and storms, coastal erosion, changes in drinking water supplies, increasing temperature);

acidification (e.g., decreases in pH, decreases in carbonate ion availability for calcifying organisms, changes in fish behavior); and nitrogen and phosphorus pollution, which could result in more profound consequences to water resources and ecosystem services. These overlapping impacts make protecting water resources in coastal areas especially challenging.

- **Indirect impacts:** Likely responses to climate change include development of alternative methods of energy and fuel production that reduce emissions of greenhouse gases, as well as finding ways to sequester carbon generated by energy production. Alternative methods of both energy production and sequestration can have impacts on water resources, including increased water use and withdrawals, potential nonpoint pollution impacts of expanded agricultural production, increased water temperatures due to discharge of process cooling waters, pollution concentration due to low flows, and effects of carbon sequestration on ground water or ocean environments.

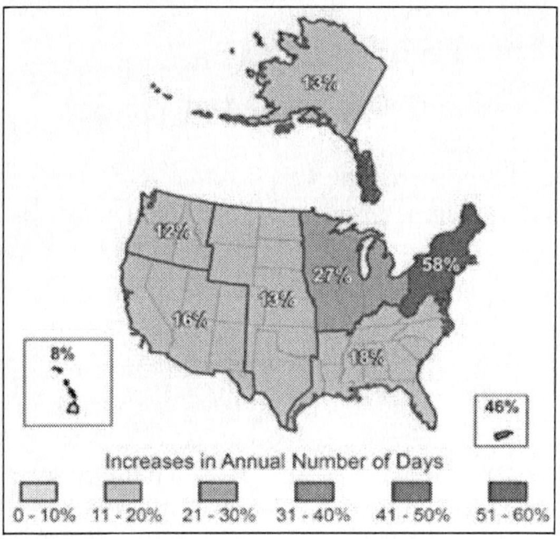

Source: U.S. Global Change Research Program (USGCRP). (2009). Global Climate Change Impacts in the United States, [Thomas R. Karl, Jerry M. Melillo, and Thomas C. Peterson, (eds.)]. New York: Cambridge University Press. Retrieved from http://www.globalchange.gov/publications/reports/scientific-assessments/us-impacts.

Figure 2. Increases in Annual Number of Days with Very Heavy Precipitation.

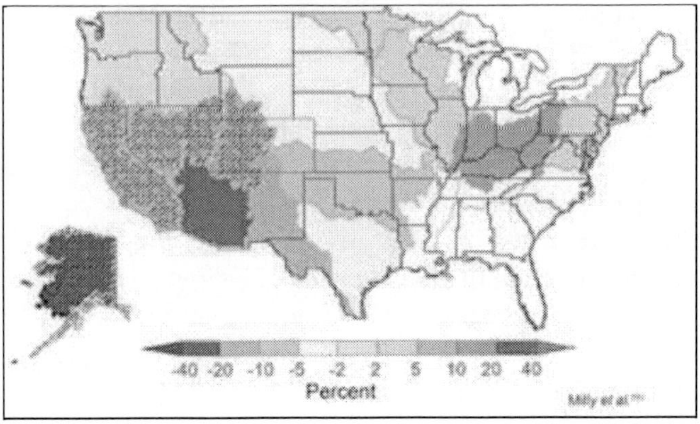

Runoff, which accumulates as streamflow, is the amount of precipitation that is not evaporated, stored as snowpack or soil moisture, or filtered down to ground water. Projected changes in median runoff for 2041–2060, relative to a 1901–1970 baseline, are mapped by water-resource region. Colors indicate percentage changes in runoff. Hatched areas indicate greater confidence due to strong agreement among model projections. White areas indicate divergence among model projections. Results are based on emissions in between the lower and higher emissions scenarios of the IPCC. Image credit: U.S. Global Change Research Program (www. globalchange.gov).

Figure 3. Projected Changes in Annual Runoff: 2041–2060.

As noted, not all near-term impacts of climate change will necessarily be disruptive and could, in some cases, provide benefits. For example, increased precipitation could improve flows supporting aquatic ecosystem health in some areas, and changing sea levels could aid submerged aquatic vegetation. (Figure 3)

Recent publications on the impacts of climate change include the June 2009 report titled *Global Climate Change Impacts in the United States*, produced by the USGCRP (formerly the U.S. Climate Change Science Program). The report reviews the scientific findings of 21 Synthesis and Assessment Products (SAPs) and builds on previous USGCRP and Intergovernmental Panel on Climate Change (IPCC) assessments. It describes both observed and expected impacts of climate change for the United States and presents regional and sectoral assessments (USGCRP, 2009a). In December 2009, EPA issued the *Endangerment and Cause or Contribute Findings for Greenhouse Gases under Section 202(a) of the Clean Air Act*. EPA relied on the major scientific assessment reports to find that greenhouse gases pose a risk to public health and welfare. Observed and projected impacts of climate change on water resources in the United States were components of the Findings (EPA, 2009a).

The National Research Council (NRC) produced a set of reports in 2010 at the request of Congress (Public Law 110-161) to study the issues associated with global climate change and provide advice on the most effective steps and strategies that can be taken to respond. The study, titled *America's Climate Choices*, resulted in five reports: *Advancing the Science of Climate Change, Limiting the Magnitude of Future Climate Change, Adapting to the Impacts of Climate Change, Informing Effective Decisions and Actions Related to Climate Change, and Synthesis for Policy Makers*, synthesizing the previous four reports (NRC, 2010a-d).

In late 2010, the NRC produced the report *Climate Stabilization Targets: Emissions, Concentrations, and Impacts over Decades to Millennia*, including an associated brochure (NRC, 2010e; NRC, 2011a). The report describes likely ranges of temperature increases during the 21st century and beyond for a given concentration of greenhouse gases in the atmosphere and associates those temperatures with likely effects on natural and human systems:

> *Scientific progress has increased confidence in the understanding of how global warming levels of 1°, 2°, 3°, 4°, 5°C, and so on, affects many aspects of the physical climate system, including regional and seasonal changes in temperature and precipitation, as well as effects on hurricanes, sea ice, snow, permafrost, sea level, and ocean acidification. Climate Stabilization Targets attempt to quantify the outcomes of different stabilization targets on the climate system, as much as is possible based on currently available scientific evidence and information (NRC, 2011a).*

The *Climate Stabilization Targets* then presents an indicative (not comprehensive) evaluation of likely impacts of each °C (1°C = 1.8°F) of warming, including, for example:

- 5–10% changes in precipitation across many regions.
- 3–10% increases in the amount of rain falling during the heaviest precipitation events.
- 5–10% changes in streamflow across many river basins.
- 15% decreases in the annually averaged extent of sea ice across the Arctic Ocean, with 25% decreases in the yearly minimum extent in September.
- 5–15% reductions in the yields of crops as currently grown.

Other effects of varying levels of warming include:

- Increases in the number of exceptionally warm summers (i.e., 9 of 10 boreal summers that are "exceptionally warm" in nearly all land areas for about 3°C of global warming, and every summer "exceptionally warm" in nearly all land areas for about 4°C, where an "exceptionally warm" summer is defined as one that is warmer than all but about one of the 20 summers in the last decades of the 20th century).
- 200–400% increases in the area burned by wildfire in parts of the western United States for 1–2°C.
- Increased coral bleaching and net erosion of coral reefs due to warming and changes in ocean acidity (pH) for carbon dioxide (CO_2) levels corresponding to about 1.5–3°C.
- Sea level rise in the range of 0.5 to 1.0 meters in 2100, in a scenario corresponding to about 3°C (plus or minus 1°C), with an associated increase in the number of people at risk from coastal flooding, as well as wetland and dryland losses.

Furthermore, the report underscores the point that "adaptation" is not a one-time event. Rather, we have entered an era of long-term continual change that must be considered by decision-makers to inform ongoing adaptation strategies. The NWP intends to continue to monitor developments in climate change and water science, including new science efforts to support and inform adaptation strategies. Notably, the USGCRP is currently conducting its third National Climate Assessment, scheduled to be final in 2013 (USGCRP, 2012). The NWP intends to incorporate into programs and activities the results of that assessment as well as of ongoing science and decision support products in the coming years.

D. The Economics of Climate Change Actions

Many of the actions we could take to adapt to climate change are actions that provide value independent of changing climate. Siting new water infrastructure in a coastal area at an elevation that is resilient to storm surge in the face of sea level rise would be beneficial even at current sea levels. Coastal wetlands are important resources for a variety of services, of which protection from sea level rise and storm surge is only one component. Sources for drinking water are already at risk; best management practices employed by water utilities and solutions encouraging water conservation and efficiency to deal with climate change impacts may also provide cost-effective relief from pressures caused by growing populations. In this sense, adaptation practices can be no- or low-regret methods to manage risk in the face of uncertainty regarding the pace and magnitude of climate change effects, provided they cost effectively address stressors in addition to the risks posed by climate change.

Quantifying the projected cost of climate change impacts with any degree of certainty is difficult due to the complexity, variability, and uncertainty in the pace, magnitude, and locally specific impacts of climate change. Likewise, it is hard to monetize the costs and benefits associated with the wide range of mitigation and adaptation opportunities available to water managers in the United States. Nevertheless, assigning a dollar value to actions and inactions related to climate change not only helps society determine its preferred level of mitigation and adaptation, but also provides a common unit of measure to compare among options, helping decision-makers determine where and how to best implement mitigation and adaptation practices. The *EPA Guidelines for Preparing Economic Analyses* (EPA, 2010b) recognizes the complexity of environmental impacts more generally, while also explaining how valuation of such impacts can benefit decision-making.

The NWP intends to monitor developments and work with partners within and outside of EPA to explore ways to characterize costs and benefits to support climate change-related decision-making. A sample of these studies follows.[4]

- Kirshen et al. (2006) quantifies the climate change impacts on water quality, water supply, and water demand, among other areas of impact, in the Boston region. For example, they estimate capital costs to account for managing lower levels of dissolved oxygen due to warmer waters to range between $30 and $39 million.
- Frederick and Schwarz (2000) look at the impact of increased flood damages and drought on the United States due to climate change, and estimate that annual average flood damages may increase from $5 billion in 1995 to $8 billion in 2030 and $18 billion in 2095.
- Dore and Burton (2001) evaluate climate adaptation costs for a variety of actions in Canada. They estimate that expanding wastewater treatment capacity in Toronto to account for more intense precipitation and other impacts could range from $533 million to $9 billion, depending on the level of risk the city is willing to accept.

- USGCRP's Global Climate Change Impacts in the United States (2009) highlights a water resources adaptation decision. Boston's Deer Island sewage treatment plant was built 1.9 feet higher to account for projected sea level rise during the facility's planned life (through 2050) to avoid future costs to build a protective wall around the plant with pumps to transport effluent over the wall.
- Neuman et al. (2010), in an EPA-supported study, evaluated the costs of sea level rise impacts to the contiguous U.S. coastline. The study found that the cost is much larger than prior estimates suggest—more than $63 billion cumulative discounted cost (at 3%) for a 27-inch rise by 2100, which corresponds to $230 billion in undiscounted cost.
- Workshop report: *Valuation Techniques and Metrics for Climate Change Impacts, Adaptation, and Mitigation Options* (NCA 2011). The goal of this workshop, convened by the interagency National Climate Assessment Task Force, was to provide a snapshot of the capabilities, readiness, and applicability of methodologies for quantitatively valuing climate impacts and adaptation.
- AWWA's recently released report, Buried No Longer, estimates that drinking water infrastructure maintenance and replacement costs will be $1 trillion from 2011–2035 for the current level of service (i.e., the cost of building climate resilience into drinking water infrastructure will be in addition to those maintenance and replacement costs).

III. Framework for a Climate Ready National Water Program

Under the clean water act, EPA and the states are directed to take a variety of actions to control pollution from point and nonpoint sources in an effort to achieve the Act's goal of attaining "water quality which provides for the protection and propagation of fish, shellfish, and wildlife and provides for recreation in and on the water." Under the SDWA, EPA promulgates national primary drinking water regulations applicable to public water systems to protect human health from drinking water contaminants. EPA's source water protection efforts aim to protect abundant and clean drinking water supplies. However, as climate change shifts hydrological patterns and increases variability outside of historic norms, including frequency, severity, and duration of drought or extreme rain events, achieving these goals will become more challenging.

A. Guiding Principles

To position the NWP as "Climate Ready," we will work with stakeholders and partners to achieve our Vision. The NWP adopted the following 10 principles that inform the development of the *2012 Strategy*. These principles are consistent with, and reinforce, the principles promulgated by the ICCATF and reflect additional principles specific to managing water resources.

1. **Integrated Water Resources Management (IWRM):** Support collaboration among state, interstate, local, tribal, and federal governments and among sectors to manage the quality and quantity of sustainable water resources within watersheds and underlying aquifers (IWRM is further discussed below).
2. **Adaptive Management:** Decisions about the future are made under some conditions of uncertainty, and adaptive management provides a method for building flexibility into policy and decision-making to manage risk and to allow for new knowledge input. Uncertainty is not necessarily a reason to defer decisions.
   The *Flexible Framework* adopted by the ICCATF (Figure 4) reflects both the evolution of climate science and the likelihood that the uncertainty regarding the timing, nature, direction, and magnitude of localized climate change impacts will continue. Because investments such as for construction of water infrastructure are capital intensive, long-lived, and require long lead times, building climate change considerations into the design of these investments is reasonable, even with some degree of uncertainty in climate projections.
3. **Collaborative Learning and Capacity Development:** Collaborate with other federal water agencies and state, interstate, tribal, and local water agencies to contribute to the development of long-range plans that account for climate change impacts. Establish partnerships to assemble and develop planning and decision support tools and the underlying datasets for climate change adaptation and mitigation.

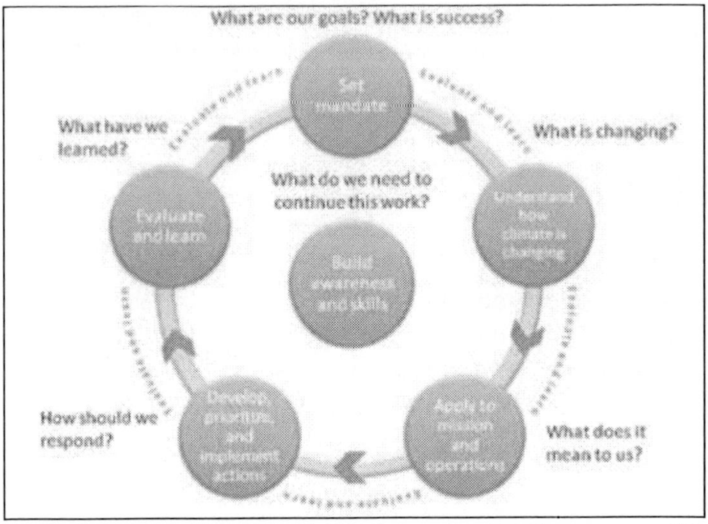

http://www.whitehouse.gov/sites/default/files/microsites/ceq/Interagency-Climate-Change-Adaptation-Progress-Report.pdf (CEQ, 2010).

Figure 4. Six-Step Approach to Climate Change Adaptation Planning.

DEALING WITH UNCERTAINTY

Although we can glean clues about the likely impacts of future climate change from recent observations and research into Earth's past, the picture is still incomplete and our predictions are uncertain. Future climate change will likely be fundamentally different from changes Earth experienced in the past because of the high temperatures that are projected, the rate of climate change, and the fact that climate change is occurring in a setting where human actions have already altered natural ecosystems in many other ways. Despite uncertainties about what the future holds, decisions can be made now. Strategies for managing ecosystems in the future will need to pay special attention to the issue of uncertainty. It will be important to make decisions based on the best currently available information, and implement them in a way that preserves the ability to make adjustments in the future as more information becomes available.

Ecological Impacts of Climate Change, [NRC, 2009]

4. **Long Term Planning (i.e., multi-decadal time horizon):** Look ahead and consider ways to reduce risk over time when making

adaptation decisions. Incorporate concepts of sustainability and non-stationarity (i.e., continual change in the hydroclimatic system outside of assumed norms) into the implementation of water programs.

5. **Energy-Water Nexus:** Saving water saves energy and vice versa. Adaptation and mitigation go hand-in-hand, and opportunities for both should be considered whenever possible. Managing the "water/energy nexus" will protect the aquatic environment while preserving freshwater resources for human uses and the economy. EPA developed a set of principles to promote these concepts to water managers and the general public (Figure 5), which are described in more detail in Appendix A.
6. **Systems & Portfolio Approach:** Design integrated and resilient solutions that address the inter-relationships among environmental, public health, social, and economic aspects of a climate change impact and that avoid unintended consequences. Incorporate diversification that includes contingency plans (emergency preparedness and response) to be implemented should adaptation actions under-perform.
7. **Cost of Inaction:** Understand the risk of inaction and its cost (i.e., the value at risk) compared to the cost of proactively adapting to projected climate change impacts. Support decision-making and express tradeoffs in terms of costs and benefits (quantified and non-quantified shortand long-term risks), as well as between action and inaction.
8. **Environmental Justice:** Account for the most vulnerable by assuring that our plans and programs consider the needs of those with a higher degree of vulnerability (e.g., children, economically disadvantaged communities, tribes, islands).
9. **Performance Evaluation:** Set clear goals against which to assess performance, and guide adaptation and refinement of program planning, policy design, and implementation. Include numeric targets where appropriate.
10. **Mainstreaming Climate Change into Core Programs:** As experience is gained and tools are developed, integrate climate change mitigation and adaptation into the NWP. Ultimately, we would no longer need a "climate change" strategy; rather, climate change would be integrated into the planning and management of our core water programs.

- **Efficiency in the use of energy and water should form the foundation of how we develop, distribute, recover, and use energy and water.**
- **The exploration, production, transmission and use of energy should have the smallest impact on water resources as possible, in terms of water quality and water quantity.**
- **The pumping, treating, distribution, use, collection, reuse and ultimate disposal of water should have the smallest impact on energy resources as possible.**
- **Wastewater treatment facilities, which treat human and animal waste, should be viewed as renewable resource recovery facilities that produce clean water, recover energy, and generate nutrients.**
- **The water and energy sectors—governments, utilities, manufacturers, and consumers—should move toward integrated energy and water management from source, production, and generation to end user.**
- **Maximize comprehensive, societal benefits.**

http://water.epa.gov/action/ energywater.cfm.

Figure 5. Principles for an Energy Water Future: The Foundation for a Sustainable America (See Appendix A for full description).

B. Integrated Water Resources Management (IWRM)

Because surface water and ground water flows across political jurisdictions, state and local government actions that are coordinated throughout watersheds and across the underlying aquifers can more successfully protect and preserve these resources than disparate actions taken piecemeal. Watershed and aquifer boundaries are the optimal organizing principle for state, interstate, tribal, and local management of fresh water to ensure these resources remain abundant and clean across the nation for current and future generations.

IWRM is a framework to holistically address current water resource issues and emerging climate change complications, such as increasing incidence of flood and drought. There are several definitions of this term, but for the purpose of this strategy, the NWP uses IWRM to describe opportunities for state, interstate, tribal, and local officials to voluntarily collaborate at watershed or aquifer scales, with support from federal agencies, to protect and preserve freshwater resources through mutually beneficial solutions. IWRM calls for intersector planning (e.g., between the energy, water, and agricultural sectors) to sustainably manage water resources. A shorthand way to think of IWRM is "one water." To be most effective, IWRM should take into account water quantity and quality, surface water and ground water, salinity of coastal estuaries, land use, floodplain management, point and nonpoint sources of pollution, green and grey infrastructure, and climate change adaptation and mitigation (EPA-R9 2011).

CASE STUDY: IWRM IN CALIFORNIA

In 2002, the Californian legislature passed the Integrated Regional Water Management (IRWM) Act and established IRWM as the framework for collaborative planning for all aspects of water resources in a region (IRWM is an example of IWRM). Between 2002 and 2006, California voters passed three Water Bonds authorizing $1.8 billion to fund competitive grants for IRWM planning and implementation. The California Department of Water Resources established guidelines for Regions to consider as they each developed their own coordination, planning and decisionmaking processes. Thus far, California has 46 active IRWM regions, covering 82% of the State. In 2011, EPA Region 9 worked with California to develop a technical guide for incorporating climate change into IRWM planning (CA, 2011a). For more information on California's program, see: www.water.ca.gov/irwm/docs/Brochures/Brochure_IRWM_020410.pdf

Strategic actions described throughout this document point to NWP efforts to work with other federal, state, interstate, and tribal agencies and other stakeholders in assembling information on the hydrologic relationships between surface water and ground water, and between water quality and quantity; developing planning support tools for water resource managers to address climate change adaptation; and building public understanding of the interaction between water use and the quality and sustainability of ground water and surface water.

The NWP intends to seek opportunities to integrate IWRM into national and regional activities and coordinate with other federal, state, interstate, tribal, and local agencies as well as with nongovernmental and private sector stakeholders to support IWRM at hydrologic scales.

IWRM is a voluntary collaboration of state, interstate, local, and tribal governments, and economic sectors, supported by federal agencies to sustainably manage the quality and quantity of water resources within watersheds and underlying aquifers.

IV. Programmatic Visions, Goals, and Strategic Actions

The national water program's over-arching vision is:

Despite the ongoing effects of climate change, the National Water Program intends to continue to achieve its mission to protect and restore our waters so that that drinking water is safe; and aquatic ecosystems sustain fish, plants and wildlife, as well as economic, recreational, and subsistence activities.

To that end, the NWP Climate Change Workgroup identified five key programmatic areas in which to apply the principles articulated above. This chapter then, is divided into five sections, each of which is organized using a three-tier framework: Vision, Goals, and Strategic Actions. Each section articulates a Vision, for which EPA may be only one of many actors. EPA intends to work collaboratively with other federal, state, interstate, tribal, and local entities to achieve each Vision.

For each Vision, we identify Goals that also reflect a long-term timeframe. The Goals articulate EPA's mission and role in achieving each Vision, and describe *what* we are trying to achieve.

Each Goal entails several Strategic Actions, which are the program building blocks to achieve the Goal. The Strategic Actions describe *how* the NWP intends to work over the next three to eight years in pursuit of our longer term Goals and Visions.

EPA intends to incorporate annual objectives into the Agency's annual budget and planning process and reflect the availability of resources and priorities. We intend to describe progress toward achieving Goals and Strategic Actions in annual reports (see the section on *Tracking Progress and Measuring Outcomes* in Chapter VI).

It is important to underscore that neither this *2012 Strategy* nor its Visions, Goals, or Strategic Actions, impose any requirements on state, tribal, or local water programs, nor do they establish any regulatory obligations on permittees or others. Rather, the *2012 Strategy* provides a comprehensive discussion of how the NWP intends, over the long term, to incorporate climate change considerations into its day-to-day activities, as appropriate and consistent with applicable statutory and regulatory authority, and in accordance with best available science and information. This document identifies areas in which the NWP intends to work with stakeholders and partners to account for and respond to the potential and actual impacts of climate change.

The five sections of this chapter are:

A. **Infrastructure** – including centralized or decentralized technologies and practices for wastewater, drinking water, and stormwater management infrastructure; Climate Ready Water Utilities; energy use and co-generation; and water supply and demand management.
B. **Watersheds and Wetlands** – including landscape strategies to protect and restore watersheds, source water areas (including ground water), and wetlands; natural infrastructure; and low impact development (LID).
C. **Coastal and Ocean Waters** – including programs for coastal wetlands and estuaries; Climate Ready Estuaries (CRE); issues associated with coastal infrastructure and coastal drinking water (e.g., sea level rise, saline intrusion); and ocean water quality, ocean habitats, and marine life.
D. **Water Quality** – including policies and programs to protect human health and ecological integrity (e.g., Water Quality Standards [WQS], Total Maximum Daily Loads [TMDLs], National Pollutant Discharge Elimination System [NPDES] permits, green infrastructure (GI) for stormwater management, and underground injection control [UIC], wellhead protection).
E. **Working With Tribes** – including how the NWP intends to use "traditional knowledge" to help guide this *2012 Strategy* and long-term implementation of adaptation measures.

A. Infrastructure

VISION: In the face of a changing climate, resilient and adaptable drinking water, wastewater and stormwater utilities (i.e., the water utility sector) ensure clean and safe water to protect the nation's public health and environment by making smart investment decisions to improve the sustainability of their infrastructure and operations, and the communities they serve, while reducing greenhouse gas emissions through greater energy efficiency.

The viability of drinking water and wastewater treatment and related infrastructure directly affects the protection of public and ecosystem health. Challenges driven by population growth, land-use change, aging infrastructure, availability of infrastructure funding, regulatory constraints, and various water quality stressors are already driving the water sector to take action. Climate change adds another dimension that will complicate these long-standing challenges for water sector operators and public officials. This chapter highlights how the NWP intends to continue assisting the water sector in achieving public health and ecosystem objectives in light of climate change and these other challenges. The recently published *Principles for an Energy Water Future* (see Appendix A) underscores many of the concepts in this section.

> "Because the perception that climate fluctuates around a stationary mean is in conflict with recently observed climate dynamics, decision makers need an approach that is responsive to changes in the likelihood of extreme outcomes as well as changes in the "average" climate ... Rather than managing the resource to maintain its past condition and state, management may need to take steps to protect the resource …or allow the resource to change as needed to adapt to climate change ... In other words the managers of these resources must work to incorporate the impact of climate change in their plans and operations."
>
> National Research Council, 2010d

Goal 1: The NWP works with the water utility sector to build the body of information and tools needed to incorporate climate change into planning and decision making to build the sector's adaptive capacity, reduce greenhouse gases, and deliver drinking water and clean water services.

This Goal highlights the objectives of the Climate Ready Water Utilities (CRWU) initiative to work with drinking water, wastewater, and stormwater utilities to advance their understanding of climate science and adaptation options. Through the CRWU program, the NWP intends to seek to expand the water sector's understanding of climate change risks and respond to the recommendations of the *Climate Ready Water Utilities: Final Report of the National Drinking Water Advisory Council* [NDWAC, 2010]. EPA's *Clean Water and Safe Drinking Water Infrastructure Sustainability Policy* [EPA, 2010c] also encourages water sector utilities to incorporate climate change considerations into their planning and operations, and supports the work of the CRWU initiative.

As recommended by the National Drinking Water Advisory Council (NDWAC) CRWU working group, climate change activities should be closely coordinated with other federal and state agencies, water sector associations, nongovernmental organizations, and tribes. CRWU activities also should be linked to other EPA programs, such as Climate Ready Estuaries (CRE) and Effective Utility Management (EUM). The EUM initiative is a collaborative partnership between EPA and major water sector associations and is based on a series of attributes of effectively managed utilities, including consideration of climate impacts. By coordinating with these and other programs, utilities can ensure that their climate change adaptation and mitigation approaches more readily address utility and community sustainability priorities through utilitywide planning, ongoing asset management and infrastructure repair and replacement, emergency response, and capacity development. Collaboration with the states through the State Revolving Fund (SRF) and other finance programs can also facilitate the consideration of climate change opportunities as states make infrastructure funding decisions.

Strategic Action 1: The CRWU program intends to work to improve access to vetted climate data and hydrological science, modeling and assessment tools. ***This action reflects the NWP's intent to incorporate climate change science and trend information into a major tool by 2015.***

Water utility officials are struggling with the number and volume of climate change studies produced by federal and state agencies, water associations, universities, and others. Concurrent with utilities moving forward to address climate change challenges, there is a strong need for continued investment in advancing the understanding of climate impacts and strategies (NDWAC, 2010). The NWP intends to continue to work with federal and state partners to improve access to hydrologic science and tools, such as trend and risk assessment tools, downscaled climate modeling, and advanced planning support models and decision support tools.

CRWU intends to refine its Climate Resilience Education and Awareness Tool (CREAT) to assist water utilities with understanding potential climate change impacts and assess their risks. CREAT allows a utility to analyze how various adaptation strategies may help reduce climate risks, enabling them to prioritize the implementation of adaptation measures. CRWU also intends to improve a searchable toolbox of resources that support all stages of the decision process, from basic climate science through integration of mitigation and adaptation into longterm planning (EPA, 2011c).

Strategic Action 2: The NWP intends to assist wastewater and drinking water treatment plants to reduce their greenhouse gas emissions and increase their long-term sustainability. The NWP intends to leverage programs such as effective utility management, sustainable asset management, and energy management, to encourage a combination of energy efficiency, co-generation and renewable energy resources.

About 80% of municipal water processing and distribution costs are for electricity, which comprises an estimated 3–4 % of national energy consumption; this percentage ranges up to 13% when residential water use is included (EPRI, 2002; EPA, 2011d). In addition, the Water Environment Research Foundation (WERF, 2010) reports that sewage typically contains 10 times the energy required to treat it, presenting an opportunity for using it as an energy source (co-generation). Becoming more energy efficient is a worthy goal for all water sector utilities and is an important step in reducing greenhouse gases and helping insulate utilities from energy costs or supply disruptions (Figure 6).

The NWP intends to continue encouraging water sector utilities to use its Energy Management Guidebook (EPA, 2008a), which uses a management systems approach to reduce energy use, along with other tools to develop sustainable energy management programs. As part of this effort, EPA intends to encourage utilities to document benefits from adopting energy management programs, such as lowering greenhouse gas emissions and operating costs.

EPA has also developed a downloadable, Excel-based Energy Use Assessment Tool that can be used by small-to medium-sized systems to conduct a utility bill and equipment analysis to assess individual baseline energy use and costs (EPA, 2012b).

The NWP intends to continue to provide information on energy-efficient and co-generation technologies in consultation with other federal agencies—principally the Department of Energy (DOE)—and continue to collaborate with the EPA's Office of Air and Radiation and other partners to improve existing tools (e.g., ENERGY STAR's Portfolio Manager) and develop new energy benchmarking and auditing tools.

Strategic Action 3: The NWP intends to work with the states and public water systems, particularly small water systems, to identify and plan for climate change challenges to drinking water safety and to assist in meeting health based drinking water standards.

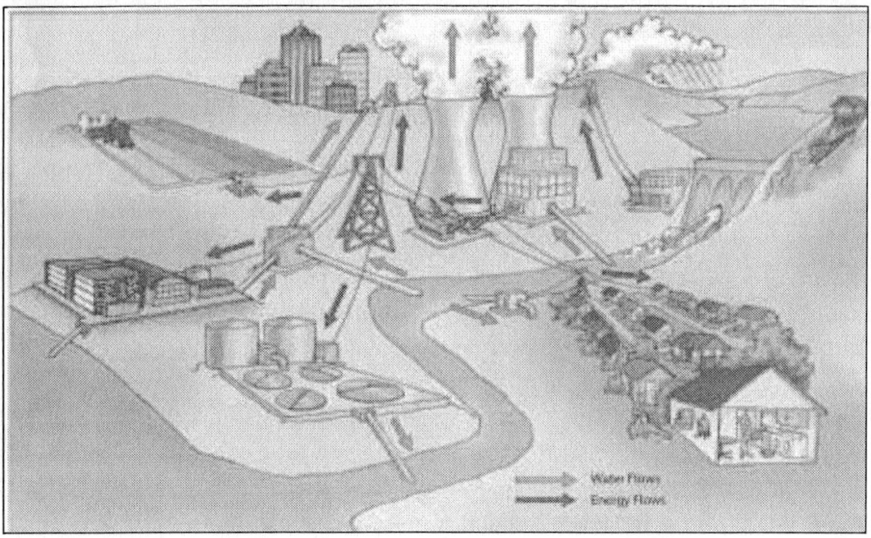

Water and energy are intimately connected. Water is used by the power generation sector for cooling, and energy is used by the water sector for pumping, treatment, and heating. Without energy there would be limited water distribution, and without water, there would be limited energy production. Image credit: U.S. Global Change Research Program (www.globalchange.gov).

Figure 6. Water and Energy Nexus.

The NWP intends to continue working to enhance partnerships with states, interstates, tribes, and others to improve water sector understanding of climate change adaptation options and identify technical assistance activities to help water systems comply with National Primary Drinking Water Regulations (NPDWRs) under changing climate conditions.

CRWU intends to focus in particular on developing tools for smaller systems. While larger utilities tend to have the resources to engage technical experts for assistance with operations, management, and decision support for climate change, smaller utilities have fewer resources. Building capacity requires providing tools and assistance tailored to smaller utilities, including tools that will prepare them to adapt to the changing climate. CRWU climate change adaptation tools will augment the capacity development efforts of the EPA drinking water program to address small system challenges affecting sustainability, compliance, and day-to-day operations. The NWP also intends to encourage partnerships between water systems to ensure they are able to avoid disruptions and consistently provide safe drinking water to their customers.

Strategic Action 4: The NWP intends to collaborate with partners to promote sustainable design approaches to ensure the long-term sustainability of infrastructure and operations. The NWP has completed *Planning for Sustainability: A Handbook for Water and Wastewater Utilities*, which provides a series of steps to help utilities voluntarily incorporate sustainability considerations into their planning. The Handbook focuses on key elements of planning, such as aligning utility sustainability goals with other community sustainability priorities in areas like housing and transportation; analyzing a range of infrastructure alternatives based on full life cycle costs, including green and natural systems; and ensuring that a financial strategy, including appropriate rate structures, is in place to fund, operate, maintain, and replace the alternatives chosen. Energy efficiency and impacts associated with climate change can be considered throughout the elements described in the handbook.

Recognizing that wastewater utilities are, in reality, resource recovery facilities, the NWP intends to work with the Water Environment Federation (WEF) and other partners to support development of an energy sustainability "roadmap." This roadmap will describe a path forward to help utilities conserve energy and become energy neutral over time. The NWP also intends to work with WEF, National Association of Clean Water Agencies, and other partners to increase public understanding of the value of biosolids as a renewable resource.

The NWP is also working with EPA's Office of Community Sustainability and three states (New York, Maryland, and California) to identify actions that can be taken to integrate the principles of the Housing and Urban Development-Department of Transportation-EPA Sustainable Communities Partnership into their Clean Water SRF programs. Options these states are considering include changes to intended use plans, project priority systems, and other funding guidance documents. Some of these changes could potentially provide incentives for projects that are energy efficient (that also help reduce greenhouse gas emissions) and/or that potentially reduce vulnerability to climate impacts. We intend to share information on the results of these pilots with other state Clean Water and Drinking Water Programs.

GOAL 2: EPA programs support IWRM in the water utility sector to sustainably manage water resources in the face of climate change.

Federal and state water resource management and protection agencies can encourage water sector utilities to establish partnerships with each other and the private sector (e.g., energy, agriculture) in the context of an IWRM framework (referred to as integrated water management in NDWAC, 2010). IWRM among water utilities and other partners can increase community resilience to climate change and expand opportunities for watershed-wide adaptive actions. The NWP, in consultation with other federal water agencies, states, interstates, and tribes, intends to consider how best to coordinate assistance to support IWRM.

Water supply management and water demand management are IWRM practices to consider, particularly where confidence in the future reliability of water supply quality or quantity is diminishing (e.g., in drought-prone, high growth, or coastal communities). The tools described below offer water sector utilities a range of methods—and there may be others—to extend their water supplies.

Many of the activities under the strategic actions for this goal can also be considered "no regrets" activities, in that they would provide benefits to utilities under current climate conditions as well as any future changes in climate.

WATER REUSE AND RECYCLING

Examples of Inter-utility IWRM in the Metropolitan Water District (MWD) of Southern California

- Orange County, California, recycles 70 million gallons per day (MGD) of sewage thru a $481 million treatment plant (NY Times, 2007) as part of a Ground Water Replenishment System (Orange County Water District, 2008).
- The City of Hemet, California, in the Eastern Municipal Water District provides recycled water to supply public parks and golf courses throughout the southland (Metropolitan Water District of Southern California, 2008).
- The Hill Canyon Water Treatment Plant (WTP) releases recycled water for agricultural irrigation under an exchange agreement between Calleguas MWD and United Water Conservation District (MWDSC, 2008).
- The Thousand Oaks Tapia WTP supplies recycled water to two MWDs for municipal and agricultural irrigation (MWDSC, 2008).

Strategic Action 5: The NWP intends to seek opportunities to better understand and promote through technical assistance the use of water supply management strategies to increase hydrologic, ecologic, public health, and economic benefit.

Water supply management can help communities build resilience when water supplies are at risk. For example, Managed Aquifer Recharge can be used to store water in aquifers for later use, and complements reuse of reclaimed wastewater to extend use, water loss control to preserve use of already treated water, and desalination to expand access to a useable resource.

Managed Aquifer Recharge: The NWP intends to work to foster research on Managed Aquifer Recharge practices that do not endanger underground sources of drinking water (USDWs). For example, Aquifer Storage and Recovery (ASR) is a process of storing water underground for future use if the injection does not endanger underground sources of drinking water. ASR is increasingly used where freshwater demand is beginning or projected to exceed supply, and use of ASR is likely to increase in drought prone areas, particularly those affected by climate change. When applied to stormwater, this practice can also reduce nonpoint source pollution of our lakes, streams, and rivers. However, the infiltration or injection of stormwater risks contamination of freshwater aquifers.

Reclamation and Reuse: The NWP intends to continue to encourage safe water reclamation and reuse. A wastewater or stormwater utility could, for example, distribute reclaimed water from a centralized treatment system for park irrigation or other uses, recognizing that additional treatment would be required for some applications. Onsite residential reuse of gray water for landscape vegetation reduces the volume of potable water delivered to the site and the volume of wastewater discharged from the centralized wastewater treatment facility. Since outdoor and non-potable water uses typically can account for more than half of all water use, this technique offers significant potential to preserve freshwater resources as well as to reduce treatment costs and energy use (EPA, 2004), and can help address increased frequency, severity, and duration of drought.

Water Loss Control: The NWP intends to provide technical assistance to reduce water loss from drinking water systems, building upon EPA's publication, *Control and Mitigation of Drinking Water Losses in Distribution Systems* (EPA, 2010d). Much of the estimated 880,000 miles of drinking water infrastructure in the United States has been in service for decades and can be a significant source of water loss. The American Water Works Association (AWWA) estimated in *Distribution System Inventory, Integrity and Water Quality* that there are close to 237,600 water line breaks per year in the United States, leading to about $2.8 billion lost in yearly revenue (EPA, 2007).

Treated water that cannot be accounted for equates to lost revenue and requires more water to be treated, which requires more energy and chemical use, which drives up operating costs. A water loss control program improves infrastructure sustainability by reducing costs and maintaining or increasing revenue. A report by the California Public Utilities Commission (CA, 2011b) found after five years of research that repairing leaks in water distribution pipes offers the highest energy savings from nine water-related strategies assessed. Water loss control also protects public health by reducing potential distribution system entry points for pathogens (EPA, 2010c).

Desalination for Potable or Nonpotable Uses: Desalination to treat marine or brackish water is becoming increasingly important in certain locations and circumstances. Several coastal communities are piloting or using desalination plants to address increasing demand driven by population growth or drought. These practices are increasing for inland sources for similar reasons or where water sources have been depleted. However, desalination is energy intensive, and there may be risks and costs associated with disposing of waste brines from the treatment. The NWP intends to monitor research developments to understand where efforts may be needed to ensure that the disposal of waste brines do not endanger underground sources of drinking water.

Strategic Action 6: The NWP intends to seek opportunities to evaluate, and provide technical assistance on, the use of water demand management strategies to increase hydrologic, ecologic, public health, and economic benefits.

Water demand management reduces consumption by providing information, technology, and incentives for consumers and industry to use less water. Water demand management calls for consumer education about the full cost of water services. To be sustainable, water utilities should be able to price water to reflect the full cost of treatment and delivery, as well as the cost of protecting water supplies. (Figure 7)

Metering and Metrics: The NWP intends to support the Water Workgroup of the ICCATF by working with other federal water agencies to develop sector-specific water-use efficiency metrics, and the NWP intends to continue seeking opportunities to assist water utilities in developing and deploying water metering technologies. Measuring water use enables development of conservation pricing as well as metrics for water-use efficiency. Service-connection metering informs customers about how much water they are using, and suppliers use metering to track water use and billing. It will also be of interest to see how the increasing use of real-time customer water use information changes customer behavior as it relates to water use.

Water Use Efficiency and WaterSense: WaterSense is an EPA-sponsored voluntary partnership to protect the future of our nation's water supply by bringing together local water utilities and governments, product manufacturers, retailers, consumers, and other stakeholders to decrease indoor and outdoor nonagricultural water use through more efficient products and practices. WaterSense helps consumers make water-efficient choices and encourages innovation in manufacturing by standardizing rigorous certification criteria that ensure product efficiency, performance, and quality (EPA, 2011f). These savings at the consumer level translate to significant direct savings in operations and maintenance costs, and indirect savings in infrastructure replacement costs, for drinking water and wastewater utilities. EPA intends to continue to develop specifications for water-efficient products; encourage water efficiency in landscape design, building operations, and codes; and educate the public on the value of water use efficiency through its WaterSense program.

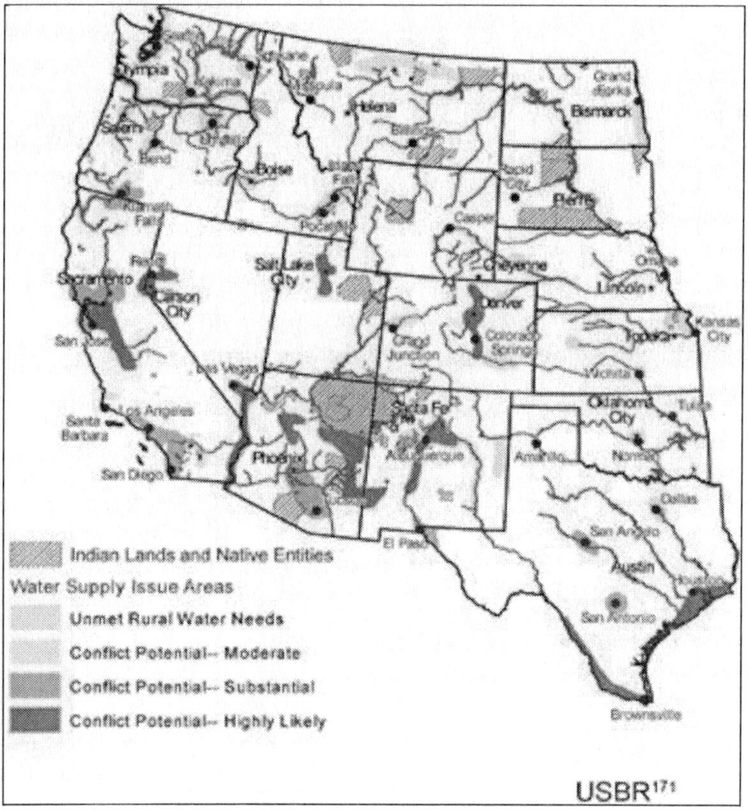

The map shows regions in the West where water supply conflicts are likely to occur by 2025 based on a combination of factors, including population trends and potential endangered species' needs for water. The red zones are where the conflicts are most likely to occur. This analysis does not factor in the effects of climate change, which is expected to exacerbate many of these alreadyidentified issues.

Image credit: U.S. Bureau of Reclamation, 2005.

Figure 7. Potential Water Supply Conflicts.

> WaterSense has helped consumers save 287 billion gallons of water and $4.7 billion in water and energy bills since 2006. By the end of 2011, WaterSense had partnered with more than 2,400 organizations and professionals, and more than 4,500 plumbing products had earned the WaterSense label (EPA, 2011e).

Water Pricing: The funding for daily operation and maintenance and long-term capital investments for drinking water and wastewater systems is typically generated through user fees. When measured as a percentage of household income, the United States pays less for water and wastewater bills than other developed countries. Because of this, there is a perception that water is readily available and water services are generally inexpensive. To meet our current and future infrastructure needs, public education on water sector system operations and costs, as well as private water conservation, is vital.

Pricing of water services should accurately reflect the true costs of providing high-quality water and wastewater services to consumers in order to both operate and maintain infrastructure and plan for upcoming repairs, rehabilitation, and replacement of that infrastructure. Drinking water and wastewater utilities should be able to price water services to reflect these costs, while also adjusting rates as needed to ensure that lower income communities can afford water and wastewater services.

There is an extensive body of knowledge on pricing water services and helping consumers learn about how pricing affects their community. EPA intends to continue to seek opportunities to work with our utility and state partners in identifying revenue templates that provide sufficient resources for infrastructure operations, maintenance, rehabilitation, and replacement, and send the right market signals about water use.

Strategic Action 7: The NWP intends to work to increase cross-sector knowledge of water supply climate challenges and develop watershed specific information to inform state, interstate, tribal, and local decision making.

It is important that state and local governments and their constituents understand the nature and extent of the water challenges they face to make decisions to address them. The NWP intends to work with federal and state science agencies and academia to develop location-specific information about climate change impacts for different sectors in each watershed and aquifer. For example, EPA is participating with other federal and state water agencies and stakeholders in planning the Department of the Interior's (DOI's) National Water Census as well as its *WaterSMART* program to promote the efficient use of water (USBR, 2011). The NWP intends to also expand its effort to collaborate with the U.S. Army Corps of Engineers as it fosters "collaborative relationships for a sustainable water resources future" (USACE, 2010a), including development of a Federal Support Toolbox to provide a common data portal to support IWRM (USACE, 2010b).

B. Watersheds and Wetlands

VISION: Watersheds are protected, maintained and restored to provide climate resilience and to preserve the ecological, social and economic benefits they provide; and the nation's wetlands are maintained and improved using integrated approaches that recognize their inherent value as well as their role in reducing the impacts of climate change.

Healthy watersheds and wetlands will be critical to climate adaptation and mitigation. This section addresses how EPA intends to protect healthy watersheds, restore impaired watersheds to enhance climate resiliency, and preserve the important functions and ecosystem services provided by the nation's wetlands, especially in the face of climate change.

Healthy watersheds and wetlands provide a host of ecological services, including water purification, ground water and surface flow regulation, wildlife habitat, flood and surge impact reduction, water temperature moderation, erosion control, and stream bank stabilization. In many cases, they also store carbon and sequester other greenhouse gases. These ecosystems already are threatened with a number of stressors, and climate change will exacerbate existing water quality and ecosystem management issues.

Protecting waters and watersheds inherently involves landscape-scale collaboration involving state, tribal, federal, and local partners. Such collaborations promote a holistic, systems approach, enabling partners to more cost-effectively reach shared goals that increase ecosystem resilience to climate change. In particular, the NWP intends to work to implement the National Fish, Wildlife and Plants Climate Adaptation Strategy (FWP, 2011), which lists seven goals (see Table 1).

Table 1. Draft National Fish, Wildlife and Plants Climate Adaptation Strategy

Goals:
Goal 1. Conserve and Connect Habitat
Goal 2. Manage Species & Habitats
Goal 3. Enhance Management Capacity
Goal 4. Support Adaptive Management
Goal 5. Increase Knowledge & Information
Goal 6. Increase Awareness & Motivate Action
Goal 7. Reduce Non-Climate Stressors

FWP, 2011. Fish, Wildlife and Plants Climate Adaptation Workgroup www.wildlife.

> "The once seemingly separable types of aquatic ecosystems are, we now know, interrelated and interdependent. We cannot expect to preserve the remaining qualities of our water resources without providing appropriate protection for the entire resource." *Tennessee Senator Howard Baker on the importance of the Clean Water Act on the Senate floor, 1977*
>
> "I ask that your marvelous natural resources be handed on unimpaired to your posterity." *Theodore Roosevelt, Sacramento, CA 1903*

The Goals and Strategic Actions in this section in particular reflect EPA's intention to implement the FWP Strategy.

GOAL 3: Identify, protect, and maintain a network of healthy watersheds and supportive habitat corridor networks across the country that provide resilience to climate change.

EPA, in partnership with others, is embarking on the Healthy Watersheds Initiative (HWI) to expand its efforts to protect healthy aquatic ecosystems using a strategic systems-based approach, prevent them from becoming impaired, and accelerate restoration (EPA, 2011g). This Initiative will greatly enhance our ability to meet the full intent and extent of the CWA 101(a) objective, "…to restore and maintain the chemical, physical, and biological integrity of the Nation's waters," and in doing so, will increase the climate resiliency of aquatic ecosystems and their watersheds. This goal would be difficult to achieve without working with our partners and their programs, such as the state-led National Fish Habitat Action Plan, the watershed protection and restoration programs of the U.S. Fish and Wildlife Service (USFWS), the National Marine Fisheries Service (NMFS), and the U.S. Forest Service

(USFS), the full suite of conservation programs administered by U.S. Department of Agriculture, the U.S. Geological Survey's (USGS) Water Smart Initiative, the Nature Conservancy's Instream Flow and North America Freshwater Conservation Programs, the U.S. Army Corps of Engineers, and others.

Integrated Water Resources Management

Successful adaptation and mitigation of climate change impacts will require a coordinated effort among all levels of government, tribes, communities, nongovernmental groups, scientific entities and the private sector that is, Integrated Water Resources Management. These voluntary partnerships will be essential to protecting and restoring watersheds, wetlands and coastal areas."

—Nancy Stoner, Acting Assistant Administrator,
EPA Office of Water, 2011

The Healthy Watersheds approach is an important component of IWRM. IWRM offers a more holistic approach to water quality protection by addressing surface water and ground water quality and quantity as one hydrologic system. As implementation of the Healthy Water-sheds approach increases our understanding of some of these relationships (e.g., hydrologic requirements of aquatic ecosystems), that knowledge will provide building blocks for the foundation of IWRM.

Strategic Action 8: The NWP intends to develop a national framework for a network of remaining healthy watersheds and aquatic ecosystems, including natural infrastructure for habitat corridors, and intends to support state and tribal efforts.

A national framework includes indicators to assess, identify, and track healthy watersheds and the success of protection measures. The NWP intends to support state and tribal efforts to conduct statewide and tribal lands integrated healthy watersheds assessments that include landscape condition (i.e., habitat corridor and floodplain connectivity and headwaters habitat intactness); hydrology; fluvial geomorphologic processes; and aquatic biology, habitat, and chemical condition. The NWP also intends to support state and tribal efforts to implement programs aimed at protecting and maintaining healthy, resilient watersheds and habitat.

The NWP intends to work with partners to develop and pilot watershed projects and management practices that improve the resilience of healthy watersheds to climate change, including the demonstration of methods that preserve and protect natural hydrology, intact active river areas (TNC, 2008), aquatic habitat corridors, natural transport of sediment, and stream geomorphology. The NWP intends to provide technical decision support to local and regional planning commissions and governments for implementing programs to protect identified watersheds in the face of climate change, consistent with the IWRM objective of the ICCATF.

Strategic Action 9: The NWP intends to collaborate with federal and other partners who focus on terrestrial ecosystems and hydrology to promote consideration of potential effects of climate change on water quality and aquatic ecosystems.

Among the multitude of services derived from intact forests are protection of water resources and sequestration of carbon. The NWP intends to continue collaborating with partner agencies (including the ICCATF Fish, Wildlife and Plants Climate Adaptation Workgroup) to support their management objectives that maximize the adaptive capacity of ecosystems (e.g., through the protection of biodiversity, functional forest groups, and keystone species, and protection against invasive species) resulting in reduced vulnerability to disturbance and associated impacts to aquatic ecosystem integrity. In particular, the NWP intends to actively support and promote appropriate forest protection efforts, afforestation (new plantings) and reforestation (replanting of deforested areas) of non-forest lands, and promote and explore partnerships with working lands, land retirement, and forestry programs within other federal agencies such as those administered by the USFS and USDA's Natural Resources Conservation Service and Farm Service Agency.

For example, to date, EPA has been working directly with USFS staff in the State and Private Forestry program to promote the use of afforestation and reforestation as a component of GI, especially as it pertains to water quality protection and stormwater management. EPA has already co-developed a draft manual describing engineered approaches to afforestation and reforestation for stormwater management and has been working through the National Arbor Day Foundation to disseminate this information to arborists, local and state forestry officials, and tree planting volunteers. EPA intends to continue working with the USFS and partners such as the National Arbor Day Foundation to support these types of outreach efforts and broaden them to address the backlog of one million acres of national forests that the USFS has identified as needing replanting.

EPA has also contributed funds to USFS staff working in the EPA Chesapeake Bay Program Office in order to start up a Web-based forestry stewardship program targeting small landowners. A geo-referenced stewardship planning tool has been developed in partnership with the Pinchot Institute and is operational for three mid-Atlantic States. The tool allows private landowners to enter information characterizing landownership in order to obtain information about available federal and state programs that encourage afforestation and reforestation while providing economic benefits. Efforts are underway to expand the coverage of this tool nationwide.

Strategic Action 10: The NWP intends to work to integrate protection of healthy watersheds throughout the NWP core programs.

Strategies that build resilience to climate change include incorporating healthy watershed protection priorities into states' continuing planning processes, promoting GI for managing stormwater, implementing the Section 404 wetlands compensatory mitigation rule, incorporating protection of healthy watersheds into funding and technical assistance programs, working with tribes, and strengthening strategic partnerships throughout EPA and the federal government, including smart growth strategies. EPA intends to encourage permitting authorities to use stormwater permits, as appropriate, to increase watershed resilience; for example, where increased use of GI or reductions in impervious cover can both address water quality issues and increase resilience to climate change. EPA intends to work with states to use the continuing planning process to develop and implement healthy watershed protection and restoration priorities.

Strategic Action 11: Increase public awareness of the role and importance of healthy watersheds in reducing the adverse impacts of climate change.

The critical ecological services watersheds and wetlands provide often go unrecognized by the public. Raising public awareness of the importance of protecting healthy watersheds will garner public support for actions needed to sustain these resources in the face of climate change.

The NWP intends to develop and implement public outreach programs emphasizing the importance of healthy watersheds, including the economic benefits of protecting and restoring watersheds, wetlands, floodplains, and riparian areas. To build support for action, the NWP intends to further articulate the climate-induced risks to aquatic ecosystems, and the associated need to enable ecosystem migration. (See for example EPA, 2011h.)

GOAL 4: Incorporate climate resilience into watershed restoration and floodplain management.

Watershed restoration and a watershed approach to floodplain management focus on reestablishing the composition, structure, pattern, and ecological processes of degraded or altered aquatic and riparian ecosystems necessary to make them sustainable, resilient, and healthy. Incorporating climate change factors into planning efforts will enable watershed strategies to be successful over the long term.

Strategic Action 12: The NWP intends to consider a means of accounting for climate change in EPA funded watershed restoration projects and encourage others funding restoration projects to take climate change and resiliency into consideration.

In partnership with other federal, state, interstate, and local water sector actors, the NWP intends to clarify and encourage implementation of existing investment flexibilities to support investments in climate resiliency in watershed restoration approaches, source water protection, GI, and joint partnerships, consistent with authorizing legislation. For example, CWA Section 319(h) grants can be used to implement nonpoint source management projects to protect and restore watersheds, including those that are vulnerable to changing land use and/ or climate change. The Section 319 grant guidance encourages partnering with other environmental programs with shared goals to leverage funding and strategically target efforts to maximize results. These partnerships are a key element to healthy watersheds protection and have the potential to be effective in meeting common goals of watershed protection across state and federal agencies.

Strategic Action 13: The NWP intends to work with federal, state, interstate, tribal, and local partners to protect and restore the natural

resources and functions of riverine and coastal floodplains as a means of building resiliency and protecting water quality.

Floodplains are among the most valuable ecosystems to society, second only to estuaries. Despite representing less than 2% of Earth's terrestrial land surface area, floodplains provide approximately 25% of all terrestrial ecosystem service benefits (Opperman, 2010). Protecting and restoring the natural resources and functions of floodplains will provide numerous environmental as well as economic benefits, such as protecting water quality, enhancing ground water recharge, and ensuring base flow of streams. Buffer areas also provide for flood attenuation, allow potential shoreline and lateral stream movement, modulate water level fluctuations, and minimize impacts on infrastructure. The NWP intends to encourage sound floodplain management, including use of nonstructural measures such as GI and LID, and work with partners to enhance the use of buffers to reduce flood losses, protect riparian ecosystems, improve water quality, and build resilience. The NWP intends to discourage use of structural measures (e.g., stream channelization and levees) whenever possible.

GOAL 5: Watershed protection practices incorporate source water protection, and vice versa, to protect and preserve drinking water supplies from the effects of climate change.

Protecting public health from contaminants in drinking water will require adapting to the impacts of climate change, which poses multiple concerns for public water systems. Warmer waters foster pathogen growth, testing the reliability of drinking water disinfection and potentially increasing costs. Increased precipitation may result in additional pollutant loadings of nutrients, pesticides, and other chemicals, further challenging drinking water treatment. Sea level rise in coastal areas puts freshwater supplies for all uses, particularly drinking water, at increasing risk. Saltwater intrusion into coastal aquifers is a problem in some areas where ground water withdrawals are outstripping recharge; increased pressure head from a higher sea level worsens this problem. As sea level rises, community drinking water intakes may end up in brackish waters as the salt front migrates up coastal rivers and streams.

Strategic Action 14: The NWP intends to encourage states to consider updating their source water delineations, assessments or protection plans to address anticipated climate change impacts.

NWP program staff intend to continue working to assure that states include protecting drinking water supplies (ground water and surface water) in watershed planning and protection programs. States should consider the feasibility and value of periodically updating their source water protection areas and protection plans in concert with state watershed plan updates to address anticipated climate change impacts. EPA and its federal partners intend to explore opportunities for providing technical assistance to states as they update their source water delineations, assessments, and protection plans to address anticipated climate change impacts.

Strategic Action 15: The NWP intends to continue to collaborate with stakeholders to increase state and local awareness of source water protection needs and opportunities and encourage inclusion of source water protection areas in local climate change adaptation initiatives.

There are many players who influence the effectiveness of source water protection at the national, state, interstate, tribal, and local levels, such as water science and regulatory agencies, water sector utility operators, local decision-makers, and nongovernmental and private sector stakeholders. Acting individually, they may affect aspects of source water protection and preservation, but collaborating on the same watersheds and aquifers increases the potential to protect and preserve those resources. The NWP intends to work to foster increased collaboration to develop decision support tools to inform deliberations at the local and watershed or aquifer scale.

GOAL 6: EPA incorporates climate change considerations into its wetlands programs, including the CWA 404 program, as appropriate.

Since 1989, the federal government as a whole has embraced a policy goal of no net loss of wetlands under the CWA Section 404 regulatory program. In addition, the program operates under a goal of a net increase in the quality and quantity of the nation's wetlands. EPA's Wetlands Program fosters effective wetlands management through strategic partnerships with states, tribes, local governments, and other partners. Key to accomplishing these goals and actions is a watershed approach to aquatic resource protection.

Section 404 of the CWA establishes a program to regulate the discharge of dredged or fill material into waters of the United States, including wetlands. Activities in waters of the United States typically regulated under this program include fill for development, water resource projects (e.g., dams and levees), infrastructure development (e.g., highways and airports), and mining projects. Section 404 requires either a permit from the U.S. Army Corps of Engineers (USACE) or an EPA-approved state program before dredged or fill material may be discharged into waters of the United States.

One basic requirement of the CWA Section 404 permitting program, as implemented by 404(b) (1) Guidelines, is that no discharge of dredged or fill material into wetlands may be permitted if a practicable alternative exists that is less damaging to the aquatic environment or if the nation's waters would be significantly degraded. Significant degradation is broadly defined in the 404(b) (1) Guidelines to include individual or cumulative impacts to human health and welfare; fish and wildlife; ecosystem diversity, productivity, and stability; and recreational, aesthetic, or economic values.

Strategic Action 16: The NWP intends to consider the effects of climate change as appropriate when making significant degradation determinations in the CWA Section 404 Wetlands Permitting and Enforcement Programs.

In light of the growing concerns regarding the adverse effects of climate change and the recognition that protecting the nation's wetlands and other aquatic resources can help mitigate these effects, EPA intends to coordinate with USACE to better understand how climate change may impact Section 404 sites and if/how the systematic consideration of climate change impacts could be incorporated into decision processes (including minimization and compensatory mitigation practices) in a scientifically and legally defensible way. EPA's Section 404 permit review process also includes determining if there would be a "substantial and unacceptable" impact to Aquatic Resources of National Importance (ARNI), as provided in Part IV of the 1992 CWA Section 404(q) Memorandum of Agreement between EPA and the Department of the Army, often called the elevation procedures. Criteria used for identifying an ARNI could potentially consider the chemical, physical, and biological importance, in light of climate change, of an aquatic resource proposed to be impacted. In partnership with USACE, the NWP also intends to consider how to incorporate the anticipated effects of climate change, as appropriate, when determining whether impacts are "unacceptable" (e.g., where discharges would result in harm to wetlands critical to storm surge reduction).

Strategic Action 17: The NWP intends to evaluate, in conjunction with relevant Federal Agencies when applicable, including USDA, USFWS, and the USACE, how wetland and stream compensation projects could be selected, designed, and sited to aid in reducing the effects of climate change.

Consistent with established regulatory policy, impacts must be compensated for "to the extent appropriate and practicable" after they are avoided and minimized to the greatest extent practicable. As an example, in order to offset permitted impacts, the Corps typically requires between 40,000 and 50,000 acres of compensatory mitigation annually. This compensation takes the form of restored, created, enhanced, and/or preserved complexes of wetlands and streams. EPA, in conjunction with the Corps, intends to consider how these wetland and stream compensation projects could be selected, designed, and sited to aid in reducing the impacts of climate change, with a focus on analyzing climate change and associated relative sea level change for coastal mitigation projects. For example, certain types of wetland mitigation projects might be encouraged in the future because of their scientifically assessed relative carbon sequestration benefits or because siting mitigation projects in coastal zones would facilitate wetland migration as sea level rises while also enhancing the natural lines of defense ("resilience") of the coastline and community and creating public green space that enhances the livability and sense of place of the community.

GOAL 7: EPA improves baseline information on wetland extent, condition, and performance to inform effective adaptation to climate change.

Baseline information on the location, extent, and quality of wetlands and aquatic resources will help to measure changes caused by climate change and other stressors. Ongoing monitoring will inform the development of predictive models and management strategies, including for climate change adaptation.

Strategic Action 18: The NWP intends to expand wetland mapping by supporting wetland mapping coalitions and training on use of the new Federal Wetland Mapping Standard.

While Agency conclusions should be informed by detailed, accurate data sources, the existing National Wetland Inventory (NWI) mapping, managed by the USFWS, is a good initial guide about potential wetlands in an area/watershed and is used extensively, including to address the effects of climate change (e.g., modeling relative sea level rise). The NWI maps were innovative when first produced, but additional work is now needed to update these maps to make them current and to better satisfy the demands for sophisticated analysis that supports effective environmental planning. Hardcopy maps are available for approximately 4/5 of the nation, and approximately half of the NWI is available online for use in geographic information system (GIS) applications. However, a significant portion of the arid Western United States has not yet been mapped.

The modernized Wetlands Mapping Standard was developed by the interagency Federal Geographic Data Committee (FGDC), in collaboration with representatives of federal agencies, states, tribes, environmental organizations, and management associations, as well as local government associations from both the wetlands and geospatial communities. The Wetland Mapping Standard was developed to improve and standardize mapping data quality in order to accelerate the rate at which the national wetlands mapping is completed and to enable realtime updates of the national wetlands data layer in the future. Using the new Standard, other groups, such as states, tribes, local governments, and nongovernmental organizations, are able to collect and upload digitally mapped data to the NWI. EPA and other federal agencies intend to train and support a range of organizations to complete the national map.

Strategic Action 19: The NWP intends to produce a statistically valid, ecological condition assessment of the nation's wetlands.

The National Wetland Condition Assessment (NWCA) will be an integrated gauge of wetland condition nationwide, summarizing the cumulative effects of federal, state, interstate, tribal, and local government and private-party actions that either degrade wetlands or protect and restore their ecological condition. The NWCA will be repeated at the national scale every five years and will incorporate those indicators that EPA identifies as most meaningful to detecting and predicting the impacts of climate change on the condition of the nation's wetlands.

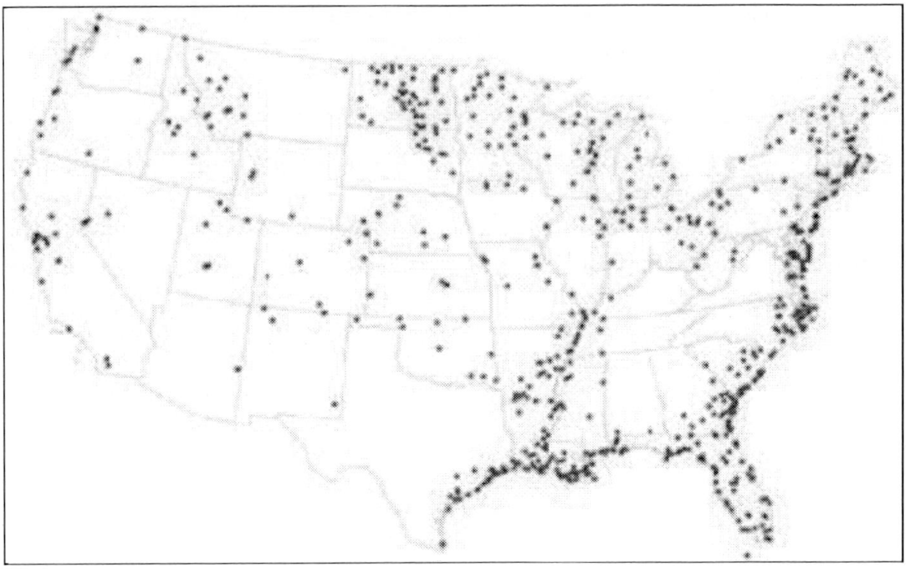

Approx. 1,000 randomly selected wetland sites sampled in 2011. National Wetland Condition Assessment, EPA, 2011n.

Figure 8. National Wetland Condition Assessment Site Locations.

EPA worked closely with the USFWS Wetlands Status and Trends program to utilize its network of analysis plots as the sampling frame for the NWCA. When these efforts are paired, we will for the first time be able to measure progress toward the national goal to increase the quantity and quality of the nation's wetlands (Figure 8).

Wetland quality or condition speaks to how wetlands differ from their "natural" state, providing an assessment of the overall ecological integrity of the resource and the relative status of wetland processes, such as the ability of a wetland to absorb nutrients. In addition, the NWCA will identify the stressors most associated with degraded wetland condition because they provide insights into the causes of declining wetland quality.

> "Wetlands are inextricably tied to water levels and changes in climatic conditions affecting water availability will greatly influence the nature and function of specific wetlands, including the type of plant and animal species within them."
>
> Secretary of the Interior Ken Salazar, announcing availability of the new wetland mapping standard. August 18, 2009

Strategic Action 20: The NWP intends to work with partners and stakeholders to develop information and tools to support long term planning and priority setting for wetland restoration projects.

Wetlands have the potential to provide added benefits for climate change adaptation as well the potential to store and sequester carbon. The NWP intends to work with partners and stakeholders to share evolving information and tools to encourage consideration of climate change in long term planning and priority setting for wetlands management strategies and sustainable restoration projects.

C. Coastal and Ocean Waters

VISION: Adverse effects of climate change along with collective stressors and unintended adverse consequences of responses to climate change have been successfully prevented or reduced in the ocean and coastal environment. Federal, tribal, state, and local agencies, organizations, and institutions are working cooperatively; and information necessary to integrate climate change considerations into ocean and coastal management is produced, readily available, and used.

Coastal and ocean environments are inextricably linked, both spatially and ecologically. This section borrows the concept of the "baseline" (a legal demarcation of ordinary low tide levels that also crosses river mouths, the opening of bays, and along the outer points of complex coastlines) to facilitate the discussion of strategies that may be more applicable to coastal environments (which we loosely define as being on the landward side of the baseline) or ocean environments (seaward of the baseline). The baseline may affect climate change strategies because of its jurisdictional implications relevant to governmental authority. However, although the terms "coastal" and "ocean" are used primarily to organize this discussion, we recognize that those domains grade into each other and that some strategies may be appropriate on both sides of the baseline.

As in other regions, coastal areas will face challenges to wetlands, watersheds, infrastructure, water quality, and drinking water. Some coastal problems, such as nonpoint source pollution and changing precipitation patterns, have the same causes and effects that are found in inland places.

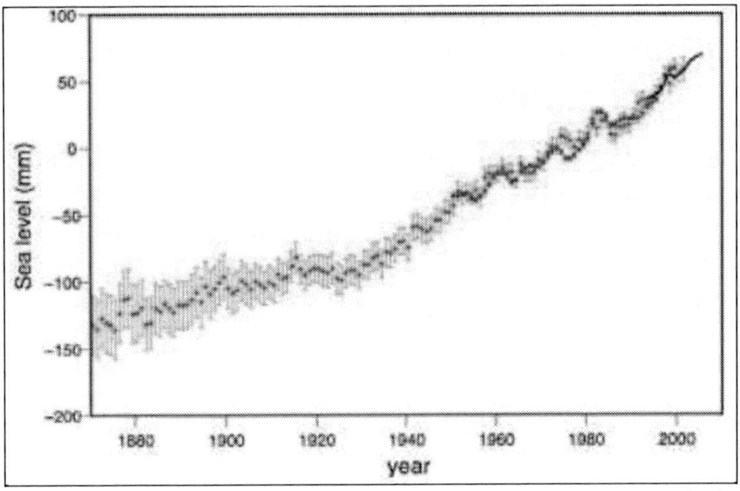

Annual, global mean sea level as determined by records of tide gauges (red and blue curves) and satellite altimetry (black curve) (NRC 2010c).

Figure 9. Sea Level Rise.

However, the ocean and coasts will experience unique impacts that the rest of the terrestrial United States will not. Sea level rise is already a multi-faceted problem that is worsening (Figure 9). Coastal wetlands and other estuarine habitats are being inundated or eroded, and many will not be able to sustain themselves as sea levels continue to rise. The potential for ocean acidification to damage the marine food chain, shellfish, and coral is another issue unique to the coastal and ocean environment. Coastal regions will also experience saltwater intrusion into ground water aquifers, the threats of rising seas to drinking water and wastewater infrastructure, and the effects of varying stream flow on estuarine salinity and ecology.

Scientific research over the last 10 years indicates that the adverse implications of ocean acidification (OA) for ocean and coastal marine ecosystems are potentially very serious because the ocean has a large capacity to absorb CO_2 from the atmosphere, and thus the resulting lowered pH levels in ocean waters can have serious cascading effects. In its 2010 report, *"Ocean Acidification: A National Strategy to Meet the Challenges of a Changing Ocean,"* the NRC (NRC, 2010f) concludes that ocean chemistry is changing at an unprecedented rate and magnitude due to human-made CO_2 emissions, and that there will be "ecological winners and losers." The report also states that "while the ultimate consequences are still unknown, there is a risk of ecosystem changes that threaten coral reefs, fisheries, protected species, and other natural resources of value to society".

Climate change impacts will in some respects be one more stressor that EPA's partners and programs will try to absorb or accommodate into their practices and portfolios. It will be important for EPA and the NWP to avert or resolve as many of the foreseeable climate adaptation problems as possible, while also preventing harm from responses to climate change that inadvertently increase vulnerability rather than reducing it.

However, in some parts of the country, such as parts of the Pacific Northwest, sea level rise is offset by coastal uplift. Such relative sea level decreases can offset absolute sea level rise and present benefits that enhance wetland preservation in coastal zones.

How others respond to the inevitable climate change impacts on coastal and ocean waters will have a large impact on EPA's ability to achieve or fulfill its mission. EPA intends to cooperate with other interested parties and work to enhance the adaptive capacity of our partners to effectively meet the coming 21st century environmental tests. We intend to work in close concert with local, state, tribal, and regional organizations and other federal partners through the wide range of existing programs and partnerships like the National Estuary Program (NEP), Large Aquatic Ecosystems (LAE), Great Waterbodies, and regional ocean groups. Working cooperatively with Canada and Mexico will also be critical if we are to be successful in near ocean environments.

Similarly, the NOC has drafted an Implementation Plan for a new, comprehensive National Ocean Policy established by Executive Order 13547 on "Stewardship of the Ocean, Our Coasts, and the Great Lakes." Following extensive stakeholder and expert input, the Plan is nearing completion in 2012. It describes a framework for federal agencies to work together to pursue common marine stewardship goals with cohesive actions, and to engage state, tribal, and local authorities; regional governance structures; nongovernmental organizations; the public; and the private sector. Table 2 lists the National Ocean Policy's nine priorities, which include Resiliency and Adaptation to Climate Change and Ocean Acidification. Upon release, the Implementation Plan will identify specific Actions and milestones for each priority, in addition to naming the federal agencies supporting those commitments. The Goals and Strategic Actions in this section reflect some of EPA's intent to implement actions under Element 6, Resiliency and Adaptation to Climate Change and Ocean Acidification, as well as other parts of the plan.

Table 2. National Ocean Policy Implementation Plan Key Elements (NOC, 2011)

1. Ecosystem-based Management
2. Inform Decisions and Improve Understanding
3. Observations, Mapping, and Infrastructure
4. Coordinate and Support
5. Regional Ecosystem Protection and Restoration
6. Resiliency and Adaptation to Climate Change and Ocean Acidification
7. Water Quality and Sustainable Practices on Land
8. Changing Conditions in the Arctic
9. Coastal and Marine Spatial Planning

GOAL 8: The NWP works collaboratively with partners so that the information and methodologies necessary to address climate change in ocean and coastal areas are collected or produced, analyzed or formatted, promoted, and easily available.

Protecting coastal and ocean environments from the adverse impacts of climate change will depend on policymakers and managers having the relevant information to make effective decisions. As the problems of climate change emerge and multiply, the need for knowledge will become even more pressing. Further, the NWP and our partners will need to know where to find the necessary information and tools. Agencies cannot afford to duplicate efforts and will need to work together to improve efficiency and leverage limited resources.

Strategic Action 21: To protect ocean and coastal areas, the NWP intends to collaborate within EPA and with other federal, tribal, and state agencies to ensure that synergy occurs whenever possible, lessons learned are transferred, federal efforts effectively help local communities and are not duplicative or working at cross-purposes.

Ensuring that lessons learned are transferred among the many partner federal agencies will maximize the utility and accessibility of new information and methodologies needed by tribal, state, and local communities to effectively prepare for climate change impacts.

Some federal agencies have already formalized cooperative mechanisms through written agreements. For example, EPA and NOAA have signed a Memorandum of Agreement to work together on climate adaptation, resilience, and smart growth efforts. In the New England region, a "Statement of Common Purpose" exists among federal agencies working together on climate change adaptation and mitigation and coastal and marine spatial planning. Similar agreements to coordinate with other federal agencies in the coastal zone, such as many DOI agencies (e.g., USGS, National Park Service, USFWS, Bureau of Ocean Energy Management), the Federal Emergency Management Agency (FEMA), USACE, DOT, and USDA (among others), would also be helpful.

Integrated Water Resources Management

As changing climates affect the decisions of water supply managers, coastal issues will be one more consideration. Managers are already balancing competing demands for in-stream ecological functions, water supply in reservoirs, water supply for downstream users, and power generation. Flows passing downstream have an effect on sediment delivery to coastal systems, the salinity structure of coastal estuaries, and how far upstream the salt front can push.

The management of coastal waters can benefit from an IWRM perspective. Issues may arise, for example, due to diversion of fresh surface water to recharge coastal aquifers, reducing flows needed for healthy coastal estuarues. Similarly, the disposal of residual brines where desalination is implemented to provide fresh water will also need attention.

Strategic Action 22: The NWP intends to work within EPA and with the U.S. Global Change Research Program and other federal, tribal, and state agencies to collect, produce, analyze, and format knowledge and information needed to protect ocean and coastal areas and make it easily available.

The NWP intends to work within EPA; with the USGCRP; and with other federal, tribal, and state agencies to produce relevant knowledge and information that informs decision-making, and to make it available in user-friendly formats through compendiums, websites, and clearinghouses. Information needed that is specific to coastal and ocean planning includes:

- Projections of relative sea level change at finer scales, including Light Detection and Ranging (LIDAR) land elevations.

- Information on ocean acidification and warming.
- Monitoring of environmental effects and system thresholds specific to the coastal and marine environments.
- Improvements in the ability to quantify real reductions of CO_2 due to salt marsh and coastal restoration.

EPA intends to continue to share similar information through portals such as ocean.data.gov and federal climate clearinghouses, such as the one under development by the USGCRP.

Please also see Strategic Action 44: Monitor climate change impacts to surface waters and ground water.

GOAL 9: Support and build networks of local, tribal, state, regional and federal collaboration partners that are knowledgeable about climate change effects and take effective adaptation measures for coastal and ocean environments through EPA's geographically targeted programs.

EPA's Large Aquatic Ecosystem Programs

- Chesapeake Bay Program
- Columbia River Basin
- Great Lakes
- Gulf of Mexico Program
- Lake Champlain Basin Program
- Long Island Sound Study
- Pacific Islands Office
- Puget Sound—Georgia Basin
- San Francisco Bay Delta Estuary
- South Florida Geographic Initiative

A primary role of the federal government will be to work within our existing networks to build adaptive capacity at the regional, state, tribal, and local levels.

Strategic Action 23: The NWP intends to work with the NWP's larger geographic programs to incorporate climate change considerations focusing on both the natural and built environments.

Geographically based programs in which EPA participates include 10 large aquatic ecosystems, Regional Ocean Partnerships, and regional planning bodies established under the National Ocean Council. The NWP intends to work to provide these key geographic programs with tools necessary to consider climate change effects in their plans and programs. EPA regional and geographic program offices and the Council of LAEs all intend to play key roles addressing climate change impacts to both the natural and built environments when making policies or decisions, and intend to work to ensure that best practices and lessons learned from local projects are widely shared.

The NWP intends to continue working with Regional Ocean Partnerships that undertake planning for resiliency. According to the Coastal States Organization's website," [t]here is an ever-growing recognition that multi-state, regional approaches are one of the most effective and efficient ways to address many of our ocean and coastal management challenges. To meet these challenges, governors around the country have voluntarily established Regional Ocean Partnerships and are working in collaboration with federal agencies, tribes, local governments, and nongovernmental and private sector stakeholders to identify shared priorities and coordinate ocean and coastal management on a regional basis. While each partnership is unique in terms of its region's issues and concerns, they all share a desire for more effective management of ocean and coastal resources. This includes balancing ecological and economic needs, and addressing climate change, through such approaches as ecosystem based management, and coastal and marine spatial planning." [CSO, 2011]

REGIONAL OCEAN AND GREAT LAKES PARTNERSHIPS

- Great Lakes Regional Collaboration
- Governors' South Atlantic Alliance
- Gulf of Mexico Alliance (GOMA)
- Mid-Atlantic Regional Council on the Ocean (MARCO)
- Northeast Regional Ocean Council (NROC)
- West Coast Governors' Agreement on Ocean Health

The NWP intends to also collaborate with the NOC's Regional Planning Bodies established under the National Ocean Policy's framework for effective Coastal and Marine Spatial Planning (CMSP) (see Executive Order 13547,

Stewardship of the Ocean, Our Coasts, and the Great Lakes). The regional planning bodies implement the framework for CMSP, leading to the eventual development of regional, coastal, and marine spatial plans that will guide and inform Agency decision-making under existing statutory authority. The NWP intends to inform the CMSP process and the development of plans to implement two priority objectives: 1) Coastal and Marine Spatial Planning (which is driven in some areas by the demand for offshore renewable energy development), and 2) Resiliency and Adaptation to Climate Change and Ocean Acidification.

Strategic Action 24: Address climate change adaptation and build stakeholder capacity when implementing NEP Comprehensive Conservation and Management Plans and through the Climate Ready Estuaries Program. Each Program intends to build its stakeholders' adaptive capacity through funding, technical assistance, and coordination.

The 28 NEPs around the country improve the quality of estuaries of national significance through community-based programs. NEPs are strategically positioned to build the adaptive capacity of stakeholders because they work directly with and within communities. In fact, many of the NEPs have specific goals in their Comprehensive Conservation and Management Plans (CCMPs) addressing climate change adaptation (Figure 10).

The Climate Ready Estuaries (CRE) program, which is jointly administered by EPA's Office of Water and Office of Air and Radiation, provides funding or direct technical assistance to help NEPs complete climate change vulnerability assessments and to build their adaptive capacity to respond to climate change. CRE intends to continue to expand its information and guidance offerings and to develop and share the lessons learned from its sponsored projects. Incorporating CRE tools and methods into the NEP base programs by mainstreaming climate change adaptation into planning documents (e.g., CCMP or annual work plans) is expected to ensure that climate change is considered by all NEPs, and helps to prepare communities to respond to climate change impacts. Many other organizations also manage coastal and ocean resources in or near NEP watersheds, so CRE intends to work collaboratively with other EPA programs (e.g., CRWU), federal agencies (e.g., NOAA's National Estuarine Research Reserves and Sea Grant, USDA's conservation planning activities), land trusts, and other nonprofit coastal organizations to build mutually supportive networks.

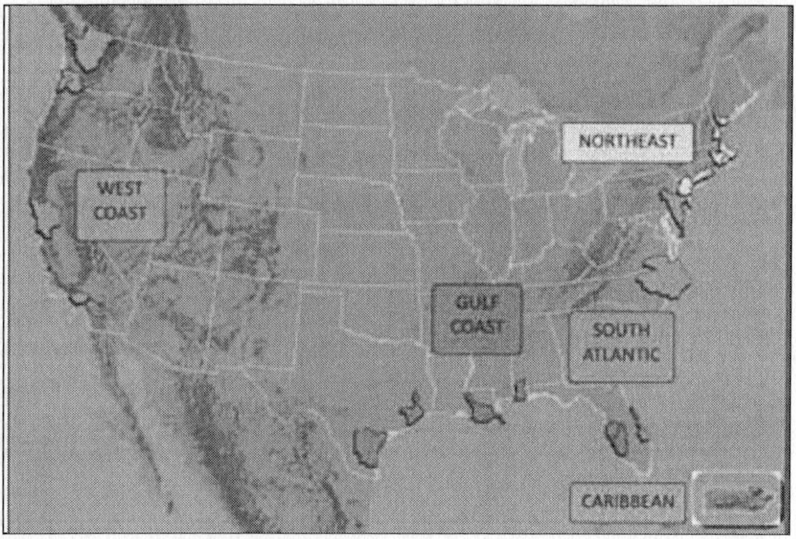

http://water.epa.gov/type/oceb/nep/index.cf

Figure 10. NEP Study Areas.

Strategic Action 25: The NWP intends to conduct outreach and education, and provide technical assistance to state and local watershed organizations and communities to build adaptive capacity in coastal areas outside the NEP and LAE programs.

All coastal areas, including regions outside NEP and LAE watersheds, should build their adaptive capacity to reduce adverse effects of climate change. The NWP can support the work of states and local watershed organizations by providing technical assistance or educational support that leverages the work of EPA's CRE and other geographic programs and partnerships. Communication will also help minimize the selection of responses to climate change that may work at cross-purposes, or have unintended adverse consequences.

GOAL 10: The NWP addresses climate driven environmental changes in coastal areas and provides that mitigation and adaptation responses to climate change are conducted in an environmentally responsible manner.

Impacts of climate change have greater consequences in coastal areas because so much of the country's population and economic infrastructure are located in those areas. Coastal areas will see greater demand for storm protection and erosion control. Strategies are needed to protect and enhance the natural environment while working toward a sustainable built environment that is prepared for climate impacts.

Coastal waters have the same potential problems with invasive species and water quality that all waters and watersheds have, in addition to marine-specific challenges such as ballast water discharges from commercial shipping. Changing precipitation patterns will affect runoff, nonpoint source pollution, and combined sewer systems, and warmer waters may foster increases in algal blooms and hypoxic conditions, decreasing the quality of waters for recreational uses such as swimming and other water sports that are extremely important in coastal areas. Warmer water will also likely worsen the already increasing occurrences of harmful algal blooms and other aspects of water quality, including the expansion in the range of many invasive species already present in U.S. waters, such as zebra mussels. Increasing temperatures in water bodies such as Lake Superior may allow organisms that have established in the other four Great Lakes to more easily establish in Superior's waters. Water bodies that were previously not receptive to invasion by many transoceanic invaders may become more habitable to those organisms.

Coastal wetlands, like all wetlands, are dependent on suitable hydrologic conditions. Climate change will severely challenge the resilience of coastal wetlands. Altered salinity from sea level rise and changing hydrometeorology will threaten coastal ecology. Geologic history and geomorphic research suggest that coastal wetlands will have a very hard time surviving at accelerated rates of sea level rise. Where salt marshes have limited sediment supplies, they will probably not be able to accrete enough material to stay above rising water level. As the intertidal zone shifts upward and landward, the area that can sustain salt marshes will shrink—in places where topography, coastal development, or insular layout prevents ecosystem shifts, marshes may disappear entirely. Some salt marshes may be able to become established upstream as the salinity changes; however, they are likely to replace freshwater tidal marshes, not to establish new wetland habitats. Both freshwater and salt marshes also are subject to changing temperature and precipitation that may affect the ability of existing species to continue to thrive.

Sea grasses are another very important aquatic resource that is vulnerable to climate change. Sea grass beds serve as critical habitat for juvenile life stages of many marine species. Most sea grass species live in a narrow strip of shallow coastal water and are extremely sensitive to changes in water clarity that control how much sunlight they receive. Warmer water, increased water depth, and turbidity from soil erosion caused by extreme precipitation and other storm events can all reduce water clarity and adversely impact the survival of sea grasses.

STORMSMARTCOASTS.ORG

...was established by the Gulf of Mexico Governor's Alliance with startup funding from NOAA and a 3-year EPA grant to expand it. Smartcoasts provides a platform for the open exchange of information among states, communities, counties, and others. A Community of Practice for Climate Change includes 100 members across the Gulf region from Sea Grant programs, NOAA, EPA, FEMA, the five Gulf of Mexico states, counties, parishes, communities, and universities.

Strategic Action 26: The NWP intends to work collaboratively to support coastal wastewater, stormwater and drinking water infrastructure owners and operators in reducing climate risks and will encourage adaptation in coastal areas.

Impacts of climate change will threaten all types of coastal infrastructure, but the water sector is particularly at risk. Sea level rise and coastal subsidence, storms and storm surge, flooding and coastal erosion, saltwater intrusion into coastal aquifers, and increasing water temperatures all threaten wastewater and drinking water treatment facilities, conveyance systems, and utility operations (See Figure 11).

EPA's CRWU and CRE programs intend to continue working together to provide coastal managers and infrastructure operators with planning support and technical assistance to help reduce climate risks and encourage adaptation. The NWP also intends to consider new approaches for ensuring that financial assistance to the water sector is used in ways that increase resiliency, reduce vulnerability, and avoid adverse unintended consequences.

Strategic Action 27: The NWP intends to work collaboratively to support climate readiness of coastal communities, including hazard mitigation, pre-disaster planning, preparedness, and recovery efforts.

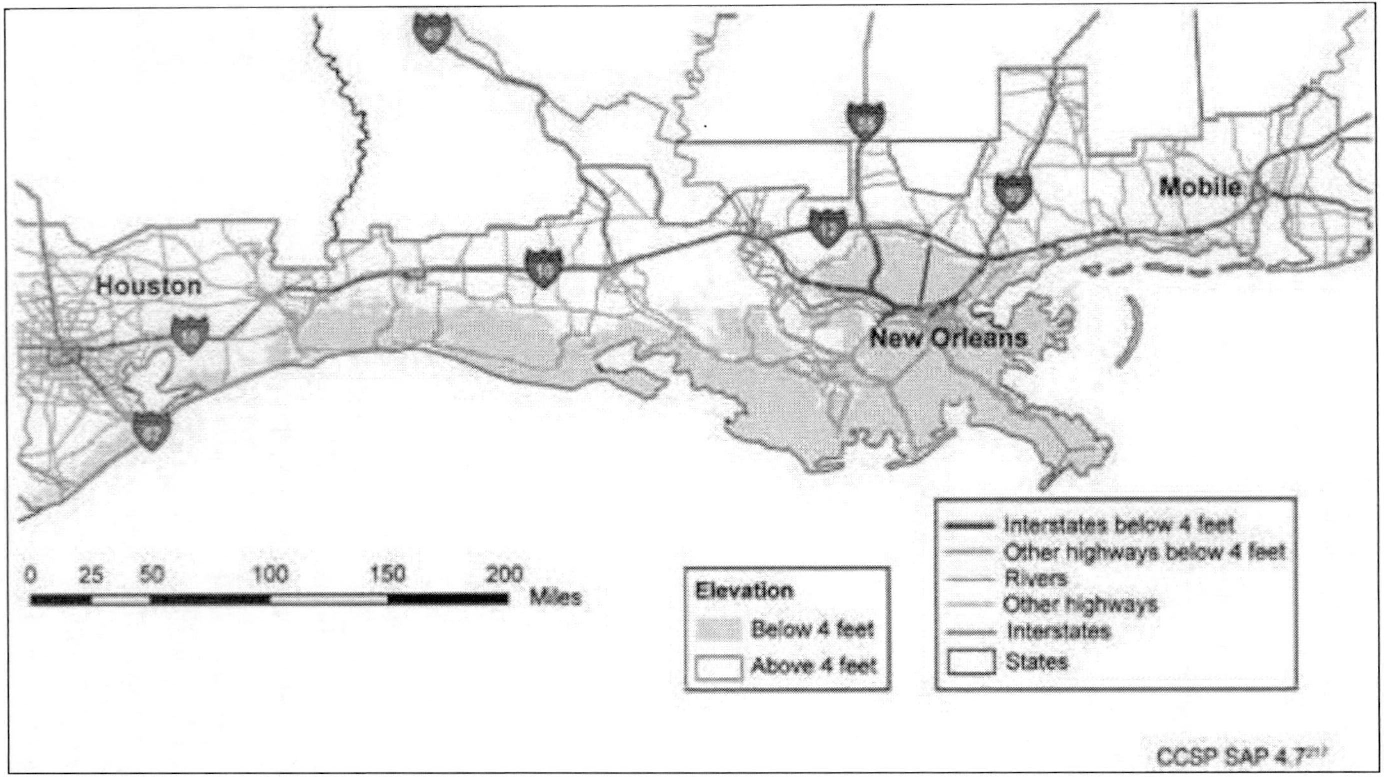

Image credit: U.S. Global Change Research Program (www.globalchange.gov).

Figure 11. Infrastructure Risk from Sea level Rise.

Climate change impacts such as sea level rise and increased storm intensity will exacerbate existing coastal hazards. Flooding, wind, waves, and storm surge that damage coastal communities can directly affect water quality, as well as damage water infrastructure.

To avoid such problems and minimize the need for emergency response, the NWP intends to work within EPA and with other federal, tribal, and state agencies to provide technical assistance to coastal communities for hazard mitigation and pre-disaster planning. After a disaster, recovery and rebuilding efforts should avoid choices that reproduce previous vulnerabilities. EPA's CRWU, CRE, and Sustainable Communities programs intend to collaborate to provide local communities with planning tools to improve resiliency to natural hazards as well as to bring other economic, environmental, and quality of life benefits. FEMA is a critical federal partner; in 2010, EPA and FEMA signed a Memorandum of Agreement that will make it easier for the two agencies to collaborate to help communities recover from disasters and better plan for future resilience, including for climate change adaptation (EPA, 2011i). The NWP also intends to coordinate with NOAA's Storm Smart Coasts program to maximize efficiencies in delivering tools and other information to local communities. These programs will assist with vulnerability analyses and help to develop and implement hazard mitigation strategies.

Local projects supported by EPA grants may be affected by climate change impacts. EPA intends to provide advice on how funding recipients can include an assessment of adaptation and mitigation measures in their planning for federally funded projects.

Strategic Action 28: The NWP intends to support preparation and response planning for a diverse array of impacts to coastal aquatic environments.

> The sea, the great unifier, is man's only hope. Now, as never before, the old phrase has a literal meaning: We are all in the same boat.
>
> — Jacques Cousteau

Coastal upland, wetland, and aquatic ecosystems and resources have evolved over centuries and millennia. They are stressed by human uses and activities and invasive species, and now face further stress from a full range of climate change impacts, including threats such as sea level rise that are unique to coastal areas.

NWP base programs and initiatives will need to be cognizant of threats to coastal water quality. While extensive expertise in restoration planning resides within EPA and at other agencies and organizations, there is a need for decision support tools to help answer challenging questions about whether restoration is viable or whether alternative strategies should be pursued in certain places. Protecting water quality and aquatic habitats such as sea grass beds may require innovative actions like ensuring that the volume and quality of freshwater inflows into estuaries are maintained. In the context of coastal change and sea level rise, decisions about coastal marshes may need to consider long-term viability and replenishment costs. The NWP intends to use existing partnerships and networks, such as the Interagency Coastal Wetlands Workgroup, Coastal America, the National Dredging Team, and other interagency planning groups, to promote the consideration of sea level change and other climate change impacts in coastal habitat restoration planning. The National and Regional Dredging Teams intend to promote the beneficial use of suitable dredged material for maintaining and restoring coastal marshes and other habitats.

In addition, "Blue Carbon" is an emerging concept that refers to the ability of aquatic ecosystems to sequester CO_2. Should emissions trading practices take hold that include Blue Carbon, the use of external funding from private sector CO_2 emissions offsets might become a useful strategy for funding restoration or creation of sustainable coastal habitats. Care should be taken, however, to ensure that other ecosystem services do not suffer if some aquatic environments are managed strictly for their ability to sequester CO_2.

GOAL 11: Protect ocean environments by incorporating shifting environmental conditions and other emerging threats into EPA programs.

Protecting the ocean environment from adverse impacts of climate change is critical to human well-being because the ocean provides food, regulates our weather, and offers numerous opportunities for renewable energy, among many other benefits. Society will also be tempted to look to the oceans for seemingly simple solutions. It is important that strategies to reduce carbon dioxide levels in the atmosphere do not impose long-term costs on ocean waters, and that the many uses of marine spaces are responsibly balanced. (Figure 12)

Strategic Action 29: The NWP intends to consider climate change impacts and associated impacts (e.g., ocean acidification, nitrogen and

phosphorus pollution) on marine water quality in its ocean management authorities, policies, and programs.

Climate change impacts to the ocean environment, including temperature increases, increased pollutant runoff, and hazardous algal blooms, as well as increases in ocean/coastal acidity, hazardous algal blooms, and spread of invasive species, add pressure to already stressed systems.

The National Coastal Conditions Report that describes the ecological and environmental conditions in U.S. coastal waters will incorporate climate change impacts into its evaluation. EPA issued a Memorandum (EPA, 2010e) that recognized the seriousness of aquatic life impacts associated with ocean acidification and described how states can move forward, where ocean acidification information exists, to address it during the CWA 303(d) listing cycle using the current 303(d) Integrated Reporting (IR) framework. Additional guidance may be necessary as improved monitoring and assessment information becomes available. If other climate change impacts on ocean environments substantially affect water quality, such as dissolved oxygen and temperature, then the NWP intends to respond to them as well (USGCRP, 2008).

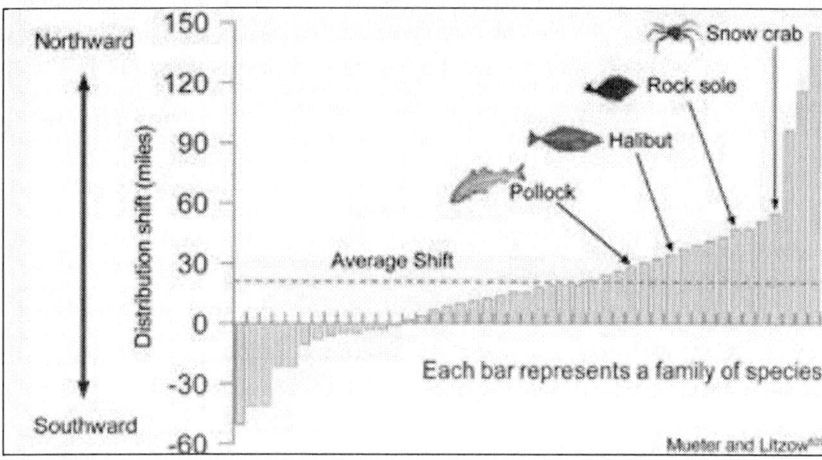

As air and water temperatures rise, marine species are moving northward, affecting fisheries, ecosystems, and coastal communitites that depend on the food source. On average, by 2006, the center of the range for the examined species moved 19 miles north of their 1982 locations.

Figure 12. Observed Northward Shift of Marine Species in the Bering Sea between the Years 1982 and 2006.

Strategic Action 30: The NWP intends to use available authorities and work with existing regional ocean governance structures, federal and state agencies, and other networks so that offshore renewable energy production does not adversely affect the marine environment.

Federal and state agencies are exploring offshore renewable energy production as a means to reduce the production of GHGs and increase energy independence. The NWP believes it is vital that the installation of renewable energy infrastructure (e.g., offshore wind turbines or wave energy systems, transmission cables, and shore-based facilities) be conducted in an environmentally responsible manner that does not result in unintended adverse consequences.

Relevant statutory authorities administered by the NWP include the National Environmental Policy Act (NEPA), the Marine Protection Research and Sanctuaries Act (MPRSA), and the CWA.

It will be particularly important to partner with and engage Regional Ocean Partnerships and EPA's geographic programs (e.g., Chesapeake Bay, Long Island Sound, Gulf of Mexico, NEPs), as well as other federal agencies, states, and tribes, and to participate in CMSP. CMSP is a comprehensive, adaptive, integrated, ecosystem-based, and transparent spatial planning process, based on sound science and intended to foster collaboration about how the ocean, coasts, and Great Lakes should be sustainably used and protected now and for future generations. Various sections of the CWA may apply to offshore energy facilities.

Strategic Action 31: The NWP intends to support the evaluation of sub-seabed sequestration of CO_2 and any proposals for ocean fertilization.

EPA intends to work with other agencies and the international community to provide technical assistance on sub-seabed carbon sequestration and coordinate with federal partners in addressing proposals for carbon sequestration in the sub-seabed or other proposals, such as potential fertilization of the ocean, including any applicable permitting that may be required under the MPRSA or the UIC program.

Carbon dioxide sequestration in sub-seabed geological formations, for example, involves separation of carbon dioxide from industrial and energy-related sources, transport to and injection into an offshore geological formation, and long-term isolation from the atmosphere. The NWP believes it is vital that the new technologies are responsibly deployed to protect the marine environment and avoid risks to coastal populations and habitats.

Strategic Action 32: The NWP intends to participate in the interagency development and implementation of federal strategies through the National Ocean Council Strategic Action Plans and the ICCATF.

Many federal agencies manage or use coastal and ocean resources to support commerce, maintain national security, and ensure environmental sustainability. The NWP intends to participate in development and implementation of federal strategies so that coastal and ocean environments are protected and are prepared for climate change adaptation and mitigation, especially through the NOC. The National Ocean Policy identifies nine priority objectives, including to "strengthen resiliency of coastal communities and marine and Great Lakes environments and their abilities to adapt to climate change impacts and ocean acidification" and "increase knowledge to continually inform and improve management and policy decisions and the capacity to respond to change and challenges." The NOC is developing a strategic action plan for this objective that will also serve as the National Action Plan (NAP) for Oceans and Coasts under the ICCATF. The NWP intends to continue to participate in writing and implementing this strategic action plan.

D. Water Quality

VISION: Our Nation's surface water, drinking water, and ground water quality are protected, and the risks of climate change to human health and the environment are diminished, through a variety of adaptation and mitigation strategies.

This section focuses on the NWP's strategy for responding to climate change impacts on water quality, using both regulatory and nonregulatory controls. Regulatory controls include WQS, TMDLs, and the NPDES, as well as drinking water regulations such as the UIC program. Nonregulatory controls include promotion of GI and LID strategies and other collaborative approaches. (Larger landscape strategies are covered in the Watersheds and

Wetlands section). This section also discusses strategies for maintaining water quality while encouraging the adoption of alternative sources of energy and fuel technologies that reduce greenhouse gas emissions.

GOAL 12: The NWP protects waters of the United States and promotes management of sustainable surface water resources under changing climate conditions.

As detailed in the *2008 Strategy*, climate change is expected to impact surface waters in several ways, affecting both human health and ecological endpoints. For example, it is projected that warmer air temperatures in many locations will heat surface water temperatures to levels that will decrease the water's ability to hold dissolved oxygen, leading to growth of harmful algal blooms and hypoxia. Warmer air temperatures may also lead to more evaporation, which could cause lower flows and higher salinity, as well as higher concentrations of other substances. Lower flows and greater salinity would likely cause an increase in impaired waters, even if actual pollutant loadings from dischargers do not increase. In many parts of the country, precipitation events are expected to become more extreme, increasing runoff with associated increases in pollutant loads, increasing variability of streamflow and associated sedimentation, and expanding flood risk.

Strategic Action 33: The NWP intends to encourage states and communities to incorporate climate change considerations into their water quality planning.

Sensitivity to impacts combined with adaptive capacity is a measure of vulnerability, and understanding vulnerability is necessary as the basis for adaptation planning. That is, the extent of climate change impacts on different ecosystems, regions, and sectors will depend not only on their sensitivity to climate change, but also on their adaptive capacity or resiliency. In order to facilitate adaptation of water programs and increase resiliency of water resources, states and tribes can use existing water quality and watershed planning programs and resources (e.g., CWA Sections 106, 604(b) and 319(h) planning funds) to conduct detailed assessments or de velop plans to increase their adaptive capacity and prioritize adaptive responses. For example, agencies or local or interstate planning organizations can use section 604(b) funds to address climate change as part of comprehensive water quality planning efforts.[5] In addition, the CWA Section 319(h) grant program can be an important resource to states for implementing nonpoint source management projects that protect vulnerable priority waters and sources of drinking and that restore impaired waters.

For more information on how NWP intends to work to protect the quality and resilience of watersheds, please see Goal 3, Strategic Action 10 in the Watersheds and Wetlands section, page 39.

INTEGRATED MUNICIPAL STORMWATER AND WASTEWATER PLANS

An integrated planning process can help define a critical path to achieving the objectives of the Clean Water Act by identifying efficiencies in implementing competing requirements that arise from separate wastewater and stormwater projects, including capital investments and operation and maintenance requirements. This approach can also lead to more sustainable and comprehensive solutions, such as green infrastructure, that improves water quality as well as supports other quality of life attributes that enhance the vitality of communities.

— EPA policy memo available at: http://cfpub. epa.gov/npdes/integratedplans.cfm (EPA, 2012d)

Strategic Action 34: The NWP intends to encourage green infrastructure and low-impact development to protect water quality and to make watersheds more resilient.

Preserving the ability of the land to absorb water helps to preserve the natural function of wetlands and watersheds while also controlling pulses of stormwater. Both GI and LID incorporate approaches to managing stormwater in a way that will reduce runoff. GI and LID management approaches and technologies use infiltration, evapotranspiration, and capture and reuse of stormwater to maintain or restore natural hydrologies (EPA, 2011j). They employ principles such as preserving and re-creating natural landscape features and minimizing imperviousness to create functional and appealing site drainage that treats stormwater as a resource rather than a waste product (EPA, 2011k). EPA is actively promoting these kinds of practices through its Green Infrastructure Strategy (EPA, 2012c), available at http://water.epa.gov/infrastructureindex.cfm.

The NWP intends to promote the use of GI and LID through tool development, stormwater permitting, outreach, and assistance programs to support states and permittees in evaluating benefits and co-benefits of GI and LID approaches. The NWP intends to consider focusing its regulatory and permitting efforts not only on new development, but also on redevelopment. This Strategic Action supports the Agency goal to incorporate climate change science and scenario information in five rulemaking processes by 2015.

Strategic Action 35: The NWP intends to promote the consideration of climate change impacts by NPDES permitting authorities.

As authorized by the CWA, the NPDES permit program reduces water pollution by regulating point sources that discharge pollutants into waters of the United States (EPA, 2009c). To help NPDES permit writers prepare for possible climate change impacts to surface waters, the NWP intends to evaluate and develop, as needed, technical tools for permit writers to improve their decision-making processes related to the impacts of climate change, such as use of precipitation and streamflow data and other data or models.

To promote water quality on a watershed scale, the NWP intends to continue to encourage the use of flexible watershed approaches, such as watershed-based permitting and water quality trading, for building surface water resiliency to climate change impacts. The NWP also intends to consider the need for, and appropriateness of, reflecting climate projections in NPDES permitting (e.g., precipitation projections).

The NWP intends to evaluate the anticipated effect of climate change on critical low-flow stream conditions, and encourage NPDES permitting authorities to incorporate revised low-flow stream estimates into NPDES permit effluent limit development where appropriate. The NWP also intends to continue to encourage NPDES permitting authorities to consider anticipated climate change impacts (e.g., warmer surface waters) when evaluating applications for 316(a) variances from thermal effluent limitations.

Strategic Action 36: The NWP intends to encourage water quality authorities to consider climate change impacts when developing wasteload and load allocations in TMDLs where appropriate.

Under Section 303(d) of the CWA, states, territories, and authorized tribes are required to develop lists of impaired waters (i.e., "the 303(d) list"). These are waters that are too polluted or otherwise degraded to meet the water quality standards set by states, territories, or authorized tribes after the implementation of effluent limitations or other pollution control requirements. The law requires jurisdictions to develop TMDLs for these waters. A TMDL is a calculation of the maximum amount of a pollutant that a water body can receive and still safely meet water quality standards (EPA, 2011l).

THE CHESAPEAKE BAY TMDL AND CLIMATE CHANGE

"EPA and USGS will work in conjunction with the states to conduct an analysis by 2017 to consider accounting for uncertainties of climate change in TMDL allocations. USGS has begun initial assessment of changes in pollution loads in the watershed under different climate and land-use scenarios. Initial results will be available in 2012 and be used to further plan assessments for TMDL allocations. Enhanced assessment will begin in 2016."

— Chesapeake Executive Order Strategy, p. 41 [CBPO, 2010]

The NWP intends to look for opportunities for states or EPA to consider potential climate change impacts when developing TMDLs. The NWP intends to explore the use of tools such as models to help states evaluate pollutant load impacts under a range of projected climatic shifts. This would be done in a way that takes into account the best available data as well as any uncertainties in the models or data.

TMDLs are developed and implemented using an adaptive management approach, in which adjustments can be made as environmental conditions, pollutant sources, or other factors change over time. Thus, as more information and tools become available, there will be opportunities to make adjustments in TMDLs to reflect climate change impacts.

Strategic Action 37: The NWP intends to identify and work to protect designated uses that are at risk from climate change impacts.

A designated use establishes the water quality goals for a specific water body and serves as the regulatory basis for establishing controls beyond technology-based requirements. The water quality standards regulations, implementing CWA section 303(c), require that states and authorized tribes specify appropriate water uses to be achieved and protected. These uses are identified by taking into consideration the use and value of the water body for public water supply; for protection and propagation of fish, shellfish, and wildlife; and for recreational, agricultural, industrial, and navigational purposes. In addition, the CWA places additional emphasis on achieving, wherever attainable, "water quality which provides for the protection and propagation of fish, shellfish, and wildlife and for recreation in and on the water" [Section 101(a)(2)]. EPA's regulation interprets and implements these provisions through requirements that WQS protect the uses specified in Section 101(a)(2) of the Act unless those uses have been shown to be unattainable.

EPA's regulations require that when removing a designated use, the state must provide an analysis (i.e., a Use Attainability Analysis [UAA]) to demonstrate that the designated use is not feasible to attain based on one of the established regulatory factors. Additionally, states are required to conduct a review of their WQS at least once every three years. As part of that triennial review, states examine whether any new information has become available for water bodies where water quality standards specify designated uses that do not include the uses specified in Section 101(a)(2) of the Act. If such new information indicates that the uses specified in Section 101(a)(2) are attainable, the state shall revise its WQS accordingly.

The water quality standards regulation specifies circumstances under which a designated use may or may not be removed or revised. If a designated use is an existing use for a particular water body, the designated use cannot be removed unless a use requiring more stringent criteria is added.

To target protective efforts, the NWP intends to identify designated uses that are important to states and tribes, necessary to meet the goals of the CWA, and vulnerable to climate change impacts. For example, recreational uses such as swimming, boating, and fishing may be affected by changes in precipitation levels, which may lead to increased impairments. Cold water fisheries may need particular consideration, since such uses may be particularly susceptible to changes in water temperature. To protect existing uses and water quality, the NWP intends to focus on implementation of antidegradation requirements, which, at a minimum, require maintenance and protection of existing uses and the level of water quality necessary to protect the existing uses.

The NWP also intends to work with stakeholders to better understand how a state could conduct Use Attainability Analysis (UAA), using the six attainability factors in EPA's current regulations, where climate change may be the primary cause of nonattainment and where impacts cannot be remedied.

Strategic Action 38: The NWP intends to clarify how states can update aquatic life water quality criteria on more regular intervals, using the best and most accurate science and data related to both changing climate conditions and how pollutants react.

Section 304(a)(1) of the CWA requires EPA to develop criteria for water quality that accurately reflect the latest scientific knowledge regarding pollutant concentrations and environmental or human health effects (EPA, 2011p). From time to time, these criteria are updated to account for advances in the science. States, tribes, and territories may adopt these criteria or other scientifically defensible criteria into their water quality standards. The NWP encourages states to update criteria using the best and most accurate science and data related to both the changing climate conditions and how pollutants react to the changing conditions on a pollutant by pollutant basis.

In addition, since climate changes will affect hydrologic conditions, the NWP intends to incorporate the best available science in an informational document to assist states and tribes that are interested in protecting aquatic life from these impacts.

GOAL 13: As the nation makes decisions to reduce greenhouse gas emissions and develop alternative sources of energy and fuel, the NWP intends to work to protect water resources from unintended adverse consequences.

Just as it takes energy to treat and distribute water supplies, it takes water to generate and produce energy and fuels. Well-designed or rehabilitated water infrastructure can reduce energy demand, and careful energy planning can reduce water demand. Using a systems approach, consolidated water infrastructure, energy, and transportation planning can directly and indirectly reduce the demand for both water and energy. While Goals 1 and 2 in the Infrastructure section of this *2012 Strategy* discuss improving the energy profile of water infrastructure, this goal identifies actions that may reduce the adverse effects of new energy technologies on water resources, consistent with the recently published *Principles for an Energy-Water Future* (see Appendix A).

Strategic Action 39: The NWP intends to continue to provide perspective on the water resource implications of new energy technologies.

Production of energy and fuel rely on access to water, and may in turn contribute to water quantity and quality problems. Further, while alternative sources of energy and fuel are important for reducing emissions of GHGs and offer a number of win-win energy choices, they too bring water resource challenges. As technologies evolve, the NWP intends to provide perspective on how the nation's energy choices affect water resources.

Strategic Action 40: EPA intends to provide assistance to states and permittees so that geologic sequestration of CO_2 is responsibly managed to protect and preserve underground sources of drinking water.

EPA finalized requirements for geologic sequestration in December 2010, under the authority of the SDWA's UIC Program (EPA, 2010e). These requirements are designed to protect underground sources of drinking water (USDWs). The rule builds on existing UIC Program requirements, with tailored requirements that address carbon dioxide injection for long-term storage to ensure that wells used for geologic sequestration are appropriately sited, constructed, tested, and monitored during and after injection (i.e., during a post injection site care period), and closed in a manner that ensures USDW protection. The NWP intends to focus on implementation of these requirements to protect USDWs.

Strategic Action 41: EPA will also continue to work with States to help them identify polluted waters, including those affected by biofuels production, and help them develop and implement Total Maximum Daily Loads (TMDLs) for those waters.

EPA finalized the Renewable Fuel Standard rulemaking in early 2010 (EPA, 2011m). The rulemaking implements a statutory provision that requires 36 billion gallons per year of biofuels be used by 2022. As the production and price of corn and other biofuel feedstocks increase, there may be impacts to both water quality and water quantity. Runoff from agricultural land carries contaminants such as fertilizers, pesticides, and sediment. More agriculture generally requires more irrigation, which increases the demand for water and the amount of water flowing directly off land and carrying pollutants into nearby water bodies. The degree to which fertilizers, pesticides, and sediment affect water quality depends on a variety of management factors, including nutrient and pesticide application rates and application methods, conservation practices and crop rotations, and acreage and intensity of tile drained lands.

Runoff from agricultural nonpoint sources is not directly controlled under the NPDES permit program. Nonpoint source pollution is addressed via state pollution control programs. These programs are supported with CWA Section 319 grant funding and include an array of regulatory and voluntary approaches depending on the state. In addition, water quality trading is a tool that can allow permitted point source facilities facing higher pollution control costs to meet their regulatory obligations by purchasing environmentally equivalent (or superior) pollution reduction credits from another source at lower cost. In some trading programs, nonpoint sources such as agricultural operations may be included in trading if pollution reductions can be sufficiently documented. EPA will also continue to work with states to help them identify polluted waters, including those affected by biofuels production, and help them develop and implement TMDLs for those waters.

Under the CWA, all point sources of pollution to a water of the United States, including ethanol plants, are required to have a permit to discharge to water bodies for both industrial process water and stormwater. NPDES permits for ethanol plants take into account the minerals, toxic pollutants, and biological oxygen demand resulting from the production process.

In order to adapt to the increased storage of biofuels, such as ethanol and biodiesel, in underground storage tanks (USTs), EPA is working with its partners to gain a better understanding of UST system materials compatibility; functionality of leak detection technologies; and the fate, transport, and remediation issues associated with biofuel releases. Unlike other fuel components, ethanol is corrosive and highly water soluble. As a result, special precautions must be taken to ensure that UST system components are both compatible and functional with ethanol blends (EPA, 2009a). EPA's ORD provides methods, models, and tools needed to remediate leaking UST sites and address fate and transport issues of leaking contaminants. EPA also proposed guidance (EPA, 2010f) that will clarify how UST owners and operators can comply with EPA's compatibility requirement, which states that owners and operators must use a system made of or lined with materials that are compatible with the substance stored in the UST system.

EPA will continue to explore these and other options for mitigation of risks related to the production and storage of biofuels, including ethanol-blended fuels, and possible impacts to water quality.

Strategic Action 42: EPA intends to provide informational materials for stakeholders to encourage consideration of alternative sources of energy and fuels that are water efficient and maintain water quality.

Alternative energy sources provide decreased reliance on fossil fuels. However, they still require access to water, and may still place added stress on water supplies. EPA intends to develop a website that consolidates EPA information on the energy/water nexus, as well as water and energy efficiency information for various sectors (forthcoming; includes EPA-OAR; EPA-R9, 2011). In order to reduce the possibility of adverse impacts to water quality and supply, EPA intends to seek opportunities and explore options to continue to develop and update outreach materials for stakeholders in concert with federal agencies such as DOE and its Renewable Energy Technology Program (DOE, 2012) and state water science agencies.

To learn more about how NWP plans to encourage energy efficiency for water utilities, please see Goal 1, Strategic Action 2 in the Infrastructure section.

Strategic Action 43: As climate change affects the operation or placement of reservoirs, the NWP intends to work with other federal agencies and EPA programs to understand the combined effects of climate change and hydropower on flows, water temperature, and water quality.

Hydropower generation is considered a renewable energy resource because the water supplying it is renewable. A hydroelectric power plant converts the downstream movement of water into electricity by directing the water, often held at a dam or reservoir, through a hydraulic turbine that is connected to a generator. Although power plants are regulated by federal and state laws to protect human health and the environment, there are a wide variety of environmental impacts associated with power generation technologies. In addition, climate change is likely to affect the amount, timing, and temperature of water used for hydropower, creating competition for water supply, affecting operational decisions, and altering the background condition of the aquatic system. The NWP intends to work with other federal agencies and programs to understand and address these combined impacts. For example, NWP could work with the DOE Wind and Water Power Program (DOE, 2011, DOE 2012) as well as with the Department of the Interior and other signatories of the Federal Hydropower Memorandum of Understanding (BOR, DOE, USACE) to further coordination and integration of hydropower and other water resource uses (BOR, 2010).

GOAL 14: The NWP intends to work to make hydrological and climate data and projections for water resource management available, when needed, in collaboration with other EPA programs and federal, state, interstate, tribal, and other partners.

Many of the NWP's programs are currently faced with a lack of sufficient data to assess national program effectiveness. Whether the data don't exist or are just not easily or publicly available differs by program, but lack of access to current data and consolidated analyses is a fundamental problem. As more climate models and vulnerability assessment tools become available, the NWP intends to work with partners from inside and outside EPA to collect, assimilate, and disseminate historic and projected information from the best sources available. The strategies in this section aim to gather, enhance, and improve access to the data that the NWP and its partners need for water resource management under changing climate conditions.

Strategic Action 44: Monitor climate change impacts to surface waters and ground water.

In order to respond to effects resulting from a changing climate, the NWP intends to understand the impacts to inland and coastal surface and ground waters. The NWP intends to support interagency monitoring networks by coordinating and collaborating with the EPA/State National Aquatic Resource Surveys (EPA, 2011n) and other agencies' monitoring programs, as well as the Federal Advisory Committee on Water Information (ACWI), to encourage them to add the ability to track and evaluate changes to water resources availability and quality using historical, reliable, long-term monitoring networks. The NWP also intends to continue to contribute to ACWI's Subcommittee on Ground Water to establish and maintain a National Ground Water Monitoring Network to describe trends in interstate and regional changes in ground water quality and availability. Further, states should understand that funding is available to assist in water quality monitoring, including surface water and ground water, under Sections 106 and 319 of the Clean Water Act. See also Strategic Action 2.

Strategic Action 45: Develop new methods for use of updated precipitation, storm frequency, and observational streamflow data, as well as methods for evaluating projected changes in low flow conditions, in collaboration with other federal agencies.

EPA intends to work to update hydrological data and methods in collaboration with federal consortia (e.g., ICCATF, the Office of Science and Technology Policy's Subcommittee on Water Availability and Quality [SWAQ], the USGCRP, the Climate Change and Water Working Group [CCAWWG], the Integrated Water Resources Science and Services [IWRSS]) and engage partners (e.g., ACWI, Water Environment Research Foundation [WERF], the Water Research Foundation [WRF]) and others to develop and standardize a process to revise precipitation, temperature, and storm event data nationwide to incorporate expected changes in commonly used data.

Of particular concern are the storm frequency, duration, and intensity estimates (e.g., 10-year, 24-hour storm events; 100-year, 24-hour storm events) and low-flow conditions in rivers and streams at the Hydrologic Unit Code 12 watershed level.

Updating precipitation records and statistical methods, and developing projections of future precipitation patterns, will enable a fundamental shift in modeling methods, which currently rely on historical data that may no longer be representative of current and future conditions. These efforts will fully consider the uncertainty inherent in predictions of the pace and magnitude of future climate-change related effects, especially at a local level.

Strategic Action 46: The NWP intends to work to enhance flow estimation using NHDPlus.

The NHDPlus is a comprehensive set of digital spatial data that encodes information about naturally occurring and constructed bodies of water, paths through which water flows, and related entities (USGS, 2011). It provides full characterization of the flow network, identification of unregulated and regulated gages and reaches, and network-based interpolation and adjustment of flows. In order to enhance flow estimates in the face of climate change, the NWP intends to support enhancements to NHDPlus as a cost-effective means of providing more accurate flow estimates for permitting, TMDLs, watershed planning, and other uses.

E. Working with Tribes

VISION: Tribes are able to preserve, adapt, and maintain the viability of their culture, traditions, natural resources, and economies in the face of a changing climate.

Native Americans have distinct cultural and spiritual connections to the water and land. The collective wisdom of elders and ancestors has allowed them to carefully use and manage the land for centuries. Changes to the earth's climate provide a new set of challenges for tribes seeking to maintain and protect their resources and the safety and health of their people.

Indian tribes are involved in protecting and restoring tens of thousands of square miles of rivers, streams, and lakes, as well as ground water in over 110,000 square miles of Indian Country in the United States. Because tribes may be regulators for water programs and water resource managers for their communities, it is important that tribes are able to provide ongoing input and participate in NWP strategies and actions on climate change. It will be important to understand and consider the impact of climate change on Native American communities and their traditional values and cultures, particularly as EPA invests in water management programs in Indian Country.

Tribes often express a holistic perspective in viewing and understanding the environment, and seek to achieve "sustainability" in their lifestyle choices, both environmentally and economically, recognizing that ultimately, it is the environment that sustains us all. Tribal recommendations to EPA include seeing the "big picture" and not compartmentalizing environmental programs into separate media to address threats from climate change.

Much of the work with tribes takes place within the EPA Regions, further described in Chapter V, Geographic Climate Regions. This section broadly describes the kinds of activities the NWP intends to pursue with tribes.

GOAL 15: The NWP incorporates climate change considerations in the implementation of its core programs for tribal nations, and collaborates with other EPA Offices and federal Agencies to work with tribes on climate change issues on a multi-media basis to build sustainability.

WORKING WITH TRIBES

Examples of EPA Adaptation Activities

Region 2 awarded a grant to the Saint Regis Mohawk Tribe to work together with all Region 2 tribal nations to discuss and design adaptation approaches during 2012.

In Region 5, the Great Lakes National Program Office is funding Great Lakes tribes to implement climate change adaptation projects and programs. Specifically, Lake Superior tribes and tribal organizations received Great Lakes Restoration Initiative (GLRI) funding through their tribal capacity grants to initiate priority climate change adaptation projects and initiatives; and tribes have been involved in the Lake Superior Sustainability Committee which is developing a Lake Superior climate change adaptation plan.

Strategic Action 47: Through formal consultation and other mechanisms, the NWP intends to ensure that the revised NWP Tribal Strategy and subsequent implementation of CWA, SDWA and other core programs incorporate climate change as a key consideration.

Principles to observe include:

- Tribes are involved in watershed-based strategies, integrated water resource management strategies, or other geographic strategies that affect tribal resources.
- Tribes participate in the development of EPA multimedia strategies for addressing climate change impacts in Indian Country.
- Actions taken are informed by and consistent with the EPA Tribal Science Council's climate change priorities and research recommendations.

Strategic Action 48: The NWP intends to incorporate adaptation into tribal funding mechanisms, and will collaborate with other EPA and federal funding programs to support sustainability and adaptation in tribal communities, to the extent appropriate and allowable by law.

Examples of actions for the NWP to pursue include:

- Provide information on the use of funding programs within the NWP to include mitigation and adaptation planning and implementation as eligible grant activities, as appropriate.
- Work with others in EPA to help clarify for tribes how funding mechanisms can be used for climate planning and implementation,

such as the Tribal General Assistance Program managed by the American Indian Environmental Office (AIEO) and Community Action for a Renewed Environment (CARE).

- Work with federal partners to coordinate tribal adaptation planning and to conduct training and education for tribal members and environmental justice communities for building adaptive capacity.

GOAL 16: Tribes have access to information on climate change that they can use to inform and engage their communities for effective decision making.

Strategic Action 49: The NWP intends to collaborate to explore and develop climate change science, information and tools for tribes to understand local climate impacts and risks to inform adaptation solutions, and will incorporate local knowledge where possible.

Examples of information requested by tribes include:

- Information on environmental conditions and long-term trends.
- Risk assessment and management tools to help identify environmental risks and inform adaptation solutions.
- Assessments of watershed conditions and impacts using peer-reviewed summaries of empirical data specific to geographic areas and water resources, to inform local action.
- Perspectives of tribal elders with historic information to inform understanding and adaptation responses.
- Management options that consider climate change factors to protect watershed resources.
- Case studies of Tribal Environmental Knowledge (TEK) incorporated into program delivery, and guidelines for incorporating TEK into science products.
- Opportunities to leverage federal resources that can provide science information to tribes.

Strategic Action 50: The NWP intends to collaborate with others to develop communication materials relevant for tribal uses and tribal audiences.

Examples of materials requested by tribes include:

- Information tailored to different climate regions.
- Information linked to tribal culture and traditions.
- Information for use in elementary, high school, and tribal college and university curricula.

V. GEOGRAPHIC CLIMATE REGIONS

A. Introduction

The USGCRP defines eight geographic regions that have broadly common climatological characteristics (USGCRP, 2009a). In evaluating the EPA water program for this revised *2012 Strategy*, we have included a discussion of particular issues by climate region. These regions are largely adopted from the USGCRP construct with a few amendments. The "Islands" Region has been broken into two distinctive Island groups (Caribbean and Pacific Islands); and a Montane Region consisting of the glaciated ranges of the Rocky Mountains, Sierra Nevada, and the Cascades was added to reflect its unique geo-graphic features and expected climate change impacts (Figure 13). Further, while the 2000 Assessment also considered "Native Peoples and Native Homelands" as a Region, we have included tribal issues in Chapter IV, Programmatic Visions, Goals, and Strategic Actions. More detailed information on climate regions can be found on the USGCRP website (USGCRP, 2012) and on EPA's main climate website (www.epa.gov/climatechange). The NWP will incorporate new information about impacts in the various climate regions as it is developed.

Several EPA Regions span multiple USGCRP regions (see Figure 13 and Table 3). Each EPA Region addresses a variety of climate impacts in their program implementation. This chapter describes strategic issues and key actions that the EPA Regions intend to focus on in the coming years, while the Regions also participate in the implementation of relevant strategic actions discussed in Chapter IV.

The federal government is working to deliver climate services not only at a national scale but also at regional scales. While similar climate characteristics can be grouped broadly within the climate regions, mitigation, and adaption efforts tend to occur at a very localized scale. EPA intends to work with other federal agencies and stakeholders to consider the spatial variability of climate change when addressing climate impacts. Examples of federal agencies working to develop localized climate-related services, and with whom each of the EPA Regional programs intends to collaborate, include:

- DOI in support of LCCs
- DOI in support of CSC
- NOAA RISAs
- Interagency/NOAA-led National Integrated Drought Information System (NIDIS)
- NOAA National Climatic Data Centers
- National Park Service Climate Friendly Parks Initiative

Further, as described in the ICCATF 2010 Progress Report (CEQ, 2010), regional offices of federal agencies have been asked to coordinate to deliver services related to climate change. As a result, an effort is underway to develop regional hubs that can provide localized assistance, where a regional adaptation coordinator can offer a single point of entry for stakeholders to access federal adaptation science and services. These partnerships will be important for EPA Regions as they work toward achieving their long-term goals, and close collaboration among these federal climate related services will be important to achieve the strategic actions described in this strategy.

Table 3. USGCRP Climate Regions and EPA Regions

Climate Regions	EPA Regions
Northeast	1, 2, 3
Southeast	3, 4, 6
Midwest	2, 5,7
Great Plains	6, 7, 8
Southwest	6, 8, 9
Pacific Northwest	8, 10
Montane	8, 9, 10
Alaska	10
Caribbean Islands	2
U.S. Pacific Islands and Territories	9

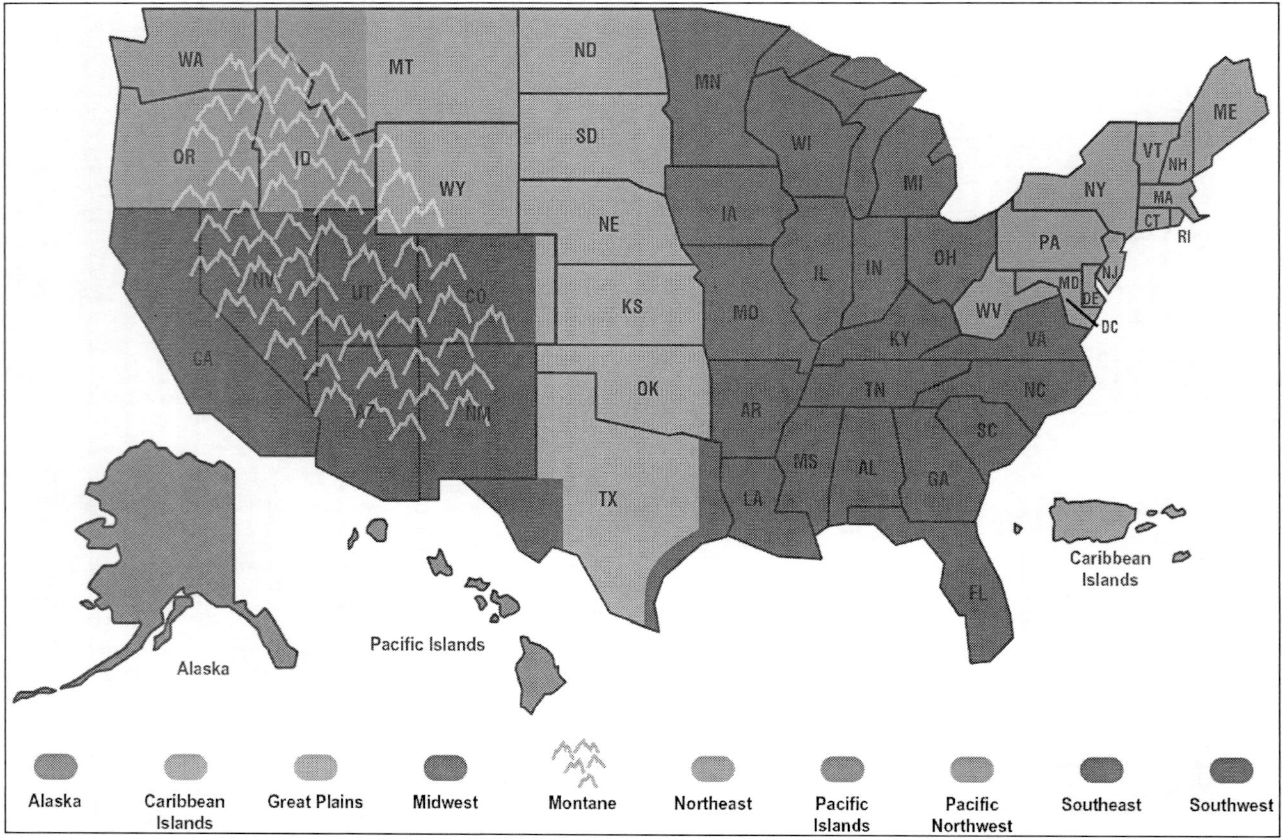

Figure 13. Climate Regions as Adapted from USGCRP.

B. Ongoing Programs Relevant to Climate Change Across All Regions

There are a number of ongoing programs and activities, described throughout this report, that are important for protecting water resources irrespective of climate change, and that are also important for both adapting to climate change impacts and reducing GHG emissions. These core programs and principles being implemented by EPA across all climate regions include:

- GI and LID.
- Water efficiency and conservation through the WaterSense program.
- Building sustainability of water and wastewater infrastructure through the CRWU program.
- Improving energy efficiency through the EUM program.
- Promoting proactive, holistic aquatic ecosystem conservation and protection through the HWI.
- Developing tools for coastal resources via the CRE program and the NEP.
- Protecting underground sources of drinking water by implementing the Geologic Sequestration rule.
- Coordinating federal funding and programs through the Partnership for Sustainable Communities between HUD, DOT, and EPA, to align infrastructure investments, such as for water, housing, or transportation, that will help reduce pollution and build resilience.

As EPA continues to develop approaches to mitigate GHG emissions and adapt to climate change, they will be adopted across the climate regions as guided by Regional priorities.

In addition, working with partners throughout the Regions will be key. One example of partners are the six interstate basin commissions that receive CWA 106 funding (Table 4) along with various other interstate commissions (ICWP, 2012).

C. EPA and Climate Regions – Goals and Strategic Actions

EPA's Regional programs provide a platform for integrating activities across media, including air, water, and land. Many of the Regions, in fact, have developed, or are developing, Regional energy and/or climate change adaptation plans or strategies. In addition, after EPA's Agency-wide adaptation plan is finalized in 2012, each EPA Region will be preparing an implementation plan providing more detail on how it will carry out the work called for in the Agency-wide plan, per the EPA Administrator's June 1, 2011, Policy Statement on Adaptation (EPA, 2011a). This section provides a synopsis of the water-related activities in the Regions and identifies long-term goals and strategic actions that EPA Regions plan to take in the coming years to build resilience at the national, state, tribal, and local levels. Links are provided to Regional websites where more information can be found.

Northeast Region

The Northeast climate region extends from West Virginia and Maryland in the south to the Canadian border in the north, and is bounded by the northern terminus of the Appalachian Mountain range to the west and the Atlantic Ocean to the east. The region includes 12 states in three EPA Regions (1, 2, and 3) and is home to 63 million people, representing 21 percent of the population of the United States. The population is concentrated along the coast, with a generally more rural interior; therefore, addressing sea level rise and other coastal issues is of particular importance.

Region 1: http://www.epa.gov/region1/climatechange/index.html
Region 2: http://www.epa.gov/region2/climate/
Region 3: http://www.epa.gov/reg3artd/globclimate/

Goal

EPA programs in the Northeast Region intend to work to make coastlines and watersheds more resilient to changes in water temperature, precipitation, and sea level.

Table 4. Interstate Organizations Receiving CWA 106 Funds

- New England Interstate Water Pollution Control Commission
- Interstate Environmental Commission (NY/NJ)
- Interstate Commission of the Potomac River Basin
- Delaware River Basin Commission
- Susquehanna River Basin Commission
- Ohio River Valley Water Sanitation Commission

Strategic Issues

- Flooding from increasingly frequent and intense rain events, as well as intense tropical storms, will tax aging infrastructure, including combined sewer systems, and adversely impact water quality.
- Dense coastal development and shoreline armoring prevents wetland migration and leads to loss of wetlands as sea level rises.
- Increases in the extent of storm surge and coastal flooding will cause erosion and property damage to the densely populated coasts. The state of New York has more than $2.3 trillion in insured coastal property (USGCRP, 2009b).
- Sea level rise may increase saltwater intrusion to coastal freshwater aquifers, resulting in water resources unusable without desalination. Increased evaporation or reduced recharge into coastal aquifers exacerbates saltwater intrusion.
- Sea level rise will lead to direct and indirect losses for the region's energy infrastructure (e.g., power plants and oil refineries located along the coast, facilities that receive oil and gas deliveries), including equipment damage from flooding or erosion. Damaged energy facilities also may be a source of pollution.
- Sea level rise, increased water temperatures, salinity distribution and circulation, changes in precipitation and freshwater runoff, and acidification will change aquatic ecosystem species composition and distribution. This will also result in the potential for new or increased prevalence of invasive species.
- Impacts from increasingly diverse types of energy development (e.g., hydraulic fracturing, biomass, land-based and offshore renewable energy development) may negatively impact the region's water resources.
- Despite the increased precipitation that most climate change models predict for the Chesapeake watershed, initial estimates of watershed models are that increases in temperature and consequent increases in evapotranspiration cause a decrease in annual river flows in the mid-Atlantic. Considering that the Baltimore, D.C., Richmond axis is the southern portion of the densely populated Boston–D.C. megalopolis, concern is warranted for securing safe and adequate drinking water supplies under climate change conditions in both the Northeast and Southeast climate regions.

Strategic Actions

In addition to promoting the core climate programs described in the introduction to this chapter, EPA is working with the New England Federal Partners group, the Mid-Atlantic Regional Council on the Ocean, and other regional networks to support the development of consistent scientific methods and robust datasets to inform long-term policy decisions on climate change vulnerability assessments and adaptation planning. This involves:

- Standardizing regional assumptions regarding future climate change impacts.
- Informing a framework for local, state, and regional decision-making that accommodates existing and emergent data sources for adaptation planning efforts.

Additionally, EPA in the Northeast Region intends to serve as a leader, coordinator, and facilitator on mitigation and adaptation activities within the region. These activities include:

- Promote water and energy efficiency at water and wastewater utilities, and encourage sustainability by promoting WaterSense, CRE, and water sustainability initiatives.
- Support the NEP and CRE programs in the development of tools and the implementation of sea level rise adaptation measures.
- Continue to engage the National Ocean Policy/NOC in addressing sea level rise adaptation and mitigation measures.
- Support emergency preparedness/response capabilities in the water sector, such as the mutual aid and assistance networks in New England.
- Promote structural and nonstructural floodplain and riparian zone management strategies that recognize that intact and well-managed watersheds are more resilient to severe storms, and absorb impacts and help balance flows over time.
- Promote adoption of GI and LID approaches through nonpoint source and stormwater management and funding programs (e.g., Municipal Separate Storm Sewer Systems [MS4] permits that include flexibility for use of LID approaches)
- Support federally recognized tribes and environmental justice populations that are already acutely impacted by water issues that may be aggravated by climate change, and may require targeted technical

assistance. For example, Region 2 awarded a grant to the Saint Regis Mohawk Tribe to work together with all Region 2 tribal nations to discuss and design adaptation approaches during 2012.

Southeast Region

The Southeast climate region extends from Virginia to the Texas border with Mexico. It includes the South Atlantic Coast, the Piedmont Coastal Plain, the Southern Appalachian Mountains, the Gulf Coast, and the southern Mississippi River Watershed. All of EPA Region 4 and parts of Regions 3 and 6 are included in the Southeast Region.

The region includes a wealth of ecological and economic resources, such as barrier islands, extensive estuaries, busy shipping ports, and important commercial and recreational fishing resources. Given the continuing population and business growth in the southeastern coastal states and the ensuing pressures on the coastal zones of this region, there are compounded pressures from decreased water supply, as well as increased flooding, sea level rise, and intense tropical storms compounded by land subsidence and heat-related stress on aquatic ecosystems and human health.

Region 3: http://www.epa.gov/reg3artd/globclimate/
Region 4: http://www.epa.gov/region4/clean_energy/index.html
Region 6: http://www.epa.gov/region6/climatechange/water.htm

Goals

Region 4 established a cross-program multimedia Energy and Climate Change Steering Committee and Workgroup that developed and is implementing an Energy and Climate Change Strategy. Similarly, Region 6 developed a Clean Energy and Climate Change Strategy through a cross-program, multimedia workgroup. These workgroups intend to work to achieve the following long-term goals:

- **Sea Level Rise:** Work with coastal states, tribes, counties, cities, and federal partners to enhance adoption of adaptive measures to lessen or avoid significant adverse effects and increase resiliency.
- **Current Data:** Update changing precipitation patterns, stream hydrology, and available water resources data and reflect them in core

water program implementation, as appropriate and taking uncertainty into consideration.

- **Water Utility Energy and Water Use Efficiency:** Promote energy and water use efficiency by working with partner utilities.
- **Geological Sequestration:** Build state programs' capacities and technical skills for implementing the Geological Sequestration Rule for Class VI wells and related permitting program.
- **Vulnerable Populations:** Work with vulnerable and historically under-represented communities to build climate change adaptation and mitigation capacities.

Strategic Issues

- Decreased water availability due to increased temperature, increased evaporation, and longer periods of time between rainfall events, coupled with an increase in societal demand, is very likely to affect many sectors of the Southeast's economy.
- Increasing evaporation and plant water loss rates alter the balance of runoff and ground water recharge, which (along with sea level rise) is likely to lead to saltwater intrusion into shallow aquifers in certain coastal areas of the Southeast.
- As sea level rises, barrier island configurations will change and coastal shorelines will retreat. Wetlands will be inundated and eroded, and low-lying areas, including some communities, will be inundated more frequently—some permanently—by the advancing sea.
- As sea level rises, temperature increases, and rainfall patterns change the salinity of estuaries, coastal wetlands, and tidal rivers, which are likely to become more variable. There will likely be longer periods of high salinities destroying coastal ecosystems or displacing them farther inland over time.
- Higher intensity and potentially more frequent storm surge flooding of coastal ecosystems and communities are likely in some low-lying areas. This concern is particularly acute along the central Gulf Coast and in south Florida and coastal North Carolina. Combined with up to 2 feet or more of sea level rise, increased storm surge is likely to result in significant human and natural resource consequences for this region.
- Hurricane intensity may increase with climate change and pose an increasingly severe risk to people, personal property, and public

infrastructure in the Southeast. Hurricanes have their greatest impact at the coastal margin where they make landfall, causing storm surge, severe beach erosion, inland flooding, and wind-related damage to both cultural and natural resources.

The warming projected for the Southeast during the next 50 to 100 years will create heatrelated stress for fish and aquatic ecosystems, and may result in a decline in dissolved oxygen in stream, lakes, and shallow aquatic habitats, leading to fish kills and loss of aquatic species diversity. Other effects of the projected increases in temperature may include more frequent outbreaks of shellfish-borne diseases in coastal waters and altered distribution of native aquatic plants and animals.

Strategic Actions

In addition to continuing to implement the ongoing climate programs described in the introduction to this chapter, Regions 3, 4, and 6 intend to support the above long-term goals through the following strategic actions:

- **Sea Level Rise**
    - Support national NEPs, focusing on the development of tools and implementing sea level adaptation measures.
    - Engage with the South Atlantic Alliance and the Gulf of Mexico Alliance, to promote resilience and reduce the impacts of (and adapt to) climate change.
    - Continue to engage with the National Ocean Policy/NOC and the Gulf Coast Ecosystem Restoration Task Force to assist in adapting to and reducing the effects of sea level rise.
    - Develop pilot regional partnerships with FEMA's Long Term Community Recovery Program to encourage pre-disaster planning and promote incorporation of sustainable, resilient reconstruction and energy management improvements into water/wastewater facilities damaged in declared disaster areas.
- **Current Data**
    - Work with EPA Headquarters, states, and tribes to incorporate changing temperatures and hydrologic data into EPA and delegated state programs.
- **Energy**
    - Recruit additional WaterSense partners, providing technical assistance and utilizing applicable grant programs.

  - Host GI training workshops and installation of GI demonstration projects. Region 4 has an ongoing project with the city of Jacksonville, Florida, to promote implementation of GI projects and principles. Region 6 is working with Dallas on a series of GI and urban heat island mitigation and adaptation projects.
  - Build internal capacity to assist water/wastewater facilities in the assessment of energy use and ways to reduce energy demands, and in identifying willing partners for which results can be measured to serve as models.
  - Develop Regional capacity for and implementation of a Regional Pilot Energy Management Program for Water/Wastewater Facilities.
- **Geologic Sequestration**
  - Host training and other technical assistance activities for states on implementation of the Geological Sequestration Rule for Class VI wells, and exercise regulatory oversight of UIC permits for CO_2 sequestration.
- **Vulnerable Populations**
  - Work with vulnerable and historically under-represented communities to ensure information, access, and attention exists for building the needed climate change adaptation and mitigation capacities.
  - Through a newly established EPA-Tribal Climate Change Network, EPA Region 6 intends to continue to work with tribal communities to provide timely and effective access to and sharing of climate change information for building mitigation and adaptation capacities in Indian Country.

Midwest Region

The Midwest's climate is shaped by the presence of the Great Lakes and the region's location in the middle of the North American continent. This location, far from the temperature-moderating effects of the oceans, experiences large seasonal swings in air temperature from hot, humid summers to cold winters. Areas from EPA Regions 5 and 7 are included in the Midwest climate region. In addition, Region 2 is connected ecologically to this climate region through the Great Lakes and the St. Lawrence Seaway. The Great Lakes are a natural resource of tremendous significance in the Midwest, containing 20 percent of the planet's fresh

surface water. Much of the region, outside of the Great Lakes Basin, drains to the Mississippi River, and as such, contributes to long-range impacts in the Gulf of Mexico. Issues of particular concern in this Region include extreme variability in precipitation and temperature, and preserving the ecological integrity of the Great Lakes.

Region 2: http://www.epa.gov/region2/climate/
Region 5: http://www.epa.gov/r5water/
http://www.epa.gov/r5climatechange/
Region 7: http://www.epa.gov/region7/water/si.htm

Goals

The long-term goals of EPA in the Midwest Region include:

- Use knowledge gained from downscaled climate models and other data to integrate climate considerations regarding precipitation into NPDES permits and long-term control plans (LTCPs), taking into account the uncertainty in model results.
- Understand vulnerability of water-related infrastructure and work with partners to increase resilience of the region's critical infrastructure to extreme storm events.
- Improve the Great Lakes community's understanding of how their ecosystems and populations will be impacted by climate change and their ability to plan and implement adaptation measures for those impacts.
- Protect ground water and surface water quality and quantity.
- Protect vulnerable populations.

Strategic Issues

- Variability in precipitation patterns will be a challenge for both drinking water and wastewater utilities and their systems. More intense rainfall can overload drainage systems and water treatment facilities, increasing the risk of waterborne diseases. This is of particular concern for combined sewer overflow (CSO) communities. Increases in such events are likely to cause greater property damage, higher insurance rates, a heavier burden on emergency management, increased cleanup and rebuilding costs, and a growing financial toll on businesses, homeowners, and insurers.

- In the summer, with increasing evaporation rates and longer periods between rainfalls, the likelihood of drought will increase, and water levels in rivers, streams, and wetlands are likely to decline. Water levels in the Great Lakes are projected to fall between one and two feet by the end of the century (USGCRP, 2009), which may result in significant lengthening of the distance to the lakeshore in many locations, impacting beaches, coastal ecosystems, dredging requirements, infrastructure, and shipping. Declining water levels in the Great Lakes will cause the migration of coastal habitats. Additionally, climate change impacts may also have profound effects on agriculture and significant resulting impacts on water quantity and quality.
- Increased water temperatures will lead to an increased risk of oxygen-poor or oxygen-free "dead zones" that kill fish and other living organisms. Warmer water and lower oxygen conditions can more readily mobilize mercury and other persistent pollutants, which is of concern for lakes with contaminated sediment. In cases where increasing quantities of contaminants are taken up in the aquatic food chain, the potential for health hazards will increase for species that eat fish from the lakes, including humans. Additionally, warming water in the Great Lakes will increase the threat of invasive species, such as zebra and quagga mussels.
- Warmer water may also exacerbate the impacts that nutrient loading has on water bodies. More intense rainfall could increase fertilizer runoff and other forms of pollutant loading to water bodies. Increased use of tile and other drainage management practices as a means to abate flooding of agricultural fields may also carry unintended consequences of increasing pollutant loading to streams and lakes.
- The Great Lakes are a bi-national resource, shared and managed jointly with Canada. Great Lakes climate change work, therefore, has a bi-national management and collaboration component.

Strategic Actions

In addition to continuing efforts in core climate change programs such as GI, WaterSense, and CRWU, specific actions to achieve the long-term goals in the Midwest Region include:

- Work with the agriculture community to consider and promote approaches such as agriculture drainage management to improve resilience and lessen water quality impacts.
- Work with states to adopt and implement EPA's Nutrient Management Framework.
- Review permit applications and issue UIC permits for CO_2 in UIC Direct Implementation states. Review primacy packages (SDWA § 1422 revision applications/GS Class VI applications) and complete the Class VI primacy approval process.
- Work with water utilities to promote energy and water efficiency.
- Engage tribes in federal climate change conversations and continue efforts to work with tribes and tribal organizations to initiate climate change adaptation projects and initiatives.
- Continue working with environmental justice populations, especially in CSO communities, to improve access to climate change information and to consider adaptation strategies.

Great Plains Region

The Great Plains climate region extends from the Dakotas and eastern half of Montana in the north to Texas in the south. On the west, it is bounded by the Rocky Mountains and the Basin and Range geographic provinces, and the central lowlands and coastal plain provinces to the east and to the south. Parts of 10 states in three EPA Regions (6, 7, and 8) are located in this vast grass-land prairie, which is home to some 9 million people, with the population expected to grow to about 14 million by 2050. The population gains will largely be in urban areas.

Region 6: http://www.epa.gov/region6/climatechange/water.htm
Region 7: http://www.epa.gov/region7/water/si.htm
Region 8: http://www.epa.gov/region8/climatechange/

Key issues in this region relate to general population growth; loss of snowpack and declining surface and ground water quality and quantity; competition for water between energy, agriculture, and public supply; and vulnerability of prairie wetlands, prairie potholes, and playa lakes.

Goals

The long-term goal of EPA in the Great Plains Region is to work to provide long-term availability and high quality of water resources and related aquatic habitat and function through:

- Water quality protection and restoration
- Water conservation and efficiency promotion
- Protection of vulnerable populations

Working specifically with partners in the agricultural sector; the renewable energy sector; and the oil, gas, and mining sectors, as well as land developers and land trusts, will be important in achieving this goal.

Strategic Issues

- General population growth, and shifts in population from the region's rural to urban centers, will continue to create demands for water storage to maintain sustainable water supplies and increase competition among water users (e.g., agricultural and municipal uses).
- Loss of snowpack in the western portion of the region will further impact water use, storage, and irrigation practices. This should be taken into consideration as infrastructure is added in the region.
- Declining surface and ground water quantity and quality, coupled with more frequent and severe droughts, will continue to exacerbate water shortages in the region.
- Unique aquatic ecosystems such as prairie wetlands, prairie potholes, and playa lakes will continue to be stressed as changes occur in ground water and surface water sources.
- Increased nonpoint source pollution (e.g., sediments, phosphorus, and nitrogen) is expected as increases in storm intensity are observed. This could result in changes to natural stream morphology and related hydrographs and could negatively impact the biological function of aquatic ecosystems.
- As in the Midwest climate region, warmer water may also exacerbate the impacts that nutrient loading has on water bodies. More intense rainfall could increase fertilizer runoff and other forms of pollutant loading to water bodies. Increased use of tile and other drainage

management practices as a means to abate flooding of agricultural fields may also carry unintended consequences of increasing pollutant loading to streams and lakes.

- Water-quality impacts will be amplified by increases in precipitation intensity and longer periods of low flow in streams.

Strategic Actions

In addition to promoting the core climate programs described in the introduction to this chapter, EPA intends to undertake the following efforts in the Great Plains Region:

- **Water Quality Protection and Restoration**
  - Work to reduce nonpoint sources of pollution to rivers and streams by leveraging the EPA's Office of Water's National Nutrients Strategy.
  - Work with partners to incorporate changing precipitation patterns, temperature, and hydrology into EPA and delegated state program decision frameworks.
  - Build geosequestration evaluation, modeling, and permitting expertise within EPA Regions through technical workshops, seminars, and related training to enhance staff capacity.
  - Exercise regulatory oversight of UIC permitting for carbon sequestration.
  - Work with states to adopt and implement EPA's Nutrient Management Framework.
- **Water Conservation and Efficiency**
  - Promote water efficiency and energy efficiency at water and wastewater utilities, and encourage sustainability by promoting WaterSense, CRE, and water sustainability initiatives such as GI initiatives workshops and related outreach efforts in major cities and along the United States–Mexico border.
- **Vulnerable Populations**
  - Work with vulnerable and historically under-represented communities to ensure the same level of information and access exists for building the needed climate change adaptation and mitigation capacities.

- Continue to work with tribal communities to provide access to climate change information, mitigation and adaptation strategies, and funding options to provide the long-term viability of natural and cultural resources that support Native American populations.

Southwest Region

The Southwest climate region includes California, Nevada, Utah, Arizona, New Mexico, and the westernmost portions of Colorado and Texas. EPA Regions 6, 8, and 9 are located in this area. To the west of the region lies the Pacific Ocean; Mexico borders the southern edge; and the Rocky Mountains border a large part of the region to the east. The population of this region, now approximately 54 million, has the fastest growth rate in the nation. The Southwest Region has multiple climatic zones, each facing somewhat different climate changes impacts. Much of the region is arid with relatively high air temperatures. Several mountain ranges as well as the Pacific Ocean influence climate and water resources in certain parts of the region. Water is stored as snowpack during the winter and released to streams in the spring and early summer, helping to meet increasing water demands. There are three major river systems: the Sacramento-San Joaquin, the Colorado, and the Rio Grande. Several huge water storage and conveyance projects also divert water from rivers for more widespread use by agriculture and growing cities. The lack of rainfall and the prospect of future droughts becoming more severe is a significant concern, especially because the Southwest continues to lead the nation in population growth.

Region 6: http://www.epa.gov/region6/climatechange/water.htm
Region 8: http://www.epa.gov/region8/climatechange/
Region 9: http://www.epa.gov/region09/climatechange/

Goals

The long-term goals of EPA in the Southwest Region are to work with federal, state, interstate, tribal, and local partners to:

- Increase the number of communities and utilities conducting climate change vulnerability assessments and implementing the resulting recommendations.

- Work with partners and stakeholders to evaluate and reduce the impacts of future drought and flooding on surface and ground water resources.
- Protect water quality and quantity to reduce stress on ecosystems.
- Address sea level rise by working with coastal states, tribes, counties, cities, and federal partners to enhance adoption of adaptive measures to lessen or avoid significant adverse effects and to increase resiliency.

Strategic Issues

- Warmer temperatures will reduce mountain snow packs, and peak spring runoff from snow melt will shift to earlier in the season, leading to and increasing the shortage of fresh water during the summer. A longer and hotter warm season will likely result in longer periods of extremely low flow and lower minimum flows in late summer. Water supply systems that have no storage or limited storage (e.g., small municipal reservoirs) may suffer seasonal shortages in summer.
- The magnitude of projected temperature increases for the Southwest, particularly when combined with urban heat island effects for major cities such as Phoenix, Albuquerque, Las Vegas, and many California cities, represents significant stresses to health, energy, and water supply in a region that already experiences very high summer temperatures.
- Reduced ground water supply due to a lack of recharge will be of concern.
- Warmer ocean temperatures may decrease productivity by stopping entrainment of deep supplies of nutrients. The resulting reductions in commercial species will need to be addressed to support continued production of fisheries and aquatic life.
- Increased frequency and altered timing of flooding will increase risks to people, ecosystems, and infrastructure. Increased flood risk is likely to result from a combination of decreased snow cover on the lower slopes of high mountains and an increased percentage of winter precipitation falling as rain and therefore running off more rapidly.
- Sea levels are rising and contributing to the loss of wetlands and infrastructure located along coastal corridors.

- The magnitude and frequency of wildfires have increased over the last 30 years, which severely impacts water quality in streams, creeks, rivers, lakes, and estuaries.

Strategic Actions

In addition to continuing to implement the ongoing climate programs described in the introduction to this chapter, EPA intends to undertake the following efforts in the Southwest:

- Encourage funding programs to fund GI, energy and water-efficient upgrades to infrastructure, and water conservation.
- Work through the California Water and Energy Project (an interagency partnership) as well as the California Financing Coordinating Committee to leverage funding to support sustainable water infrastructure and water-use efficiency projects.
- Continue to provide funding for tribal sustainable water infrastructure projects in coordination with the Indian Health Services.
- Build partners' and stakeholders' understanding of, and the capacity to respond to, risks of climate change and water.
- Work with states and local governments to expand water sources, storage, and recovery options (e.g., aquifer storage and recharge, water reuse, desalination) for areas experiencing snow pack loss and drought.

Pacific Northwest Region

The Pacific Northwest climate region includes Washington, Oregon, Idaho, and the western third of Montana. It is bounded by the Pacific Ocean on the west and the Rocky Mountains on the east and includes EPA Region 10 and part of Region 8. Canada borders the region to the north. Of primary concern are current impacts related to changes in snowpack, stream flows, sea level, forests, and other important aspects of life in the Northwest, with more severe impacts expected over the coming decades in response to continued and more rapid climate change.

Goals

The long-term goals of EPA in the Northwest Region are to work with federal, state, interstate, tribal, and local partners to increase sustainability and reduce vulnerability of communities and infrastructure, including by conserving water and increasing infiltration, and to partner with other federal agencies and the regional CSC to coordinate and leverage climate research and other activities.

Region 8: http://www.epa.gov/region8/climatechange/
Region 10: http://yosemite.epa.gov/R10/ECOCOMM.NSF/climate+change/cc

Strategic Issues

- Salmon and other coldwater species will experience additional stresses as a result of rising water temperatures and declining summer streamflows.
- Sea level rise along vulnerable coastlines will result in increased erosion and loss of land.
- Declining springtime snowpack will lead to reduced summer streamflows, straining water availability for all uses.
- Increased insect outbreaks, wildfires, and changing species composition in forests will pose challenges for ecosystems and the forest products industry.
- Water supplies will become increasingly scarce, calling for tradeoffs among competing uses, and potentially leading to conflict.
- Increased frequency of flooding will increase risk to people, ecosystems, and infrastructure.
- Projected heavier winter rainfall may cause an increase in saturated soils and therefore an increased number of landslides, particularly where there have been intensive development or forest practices on unstable slopes.
- Agriculture, ranching, and natural lands—already under pressure due to an increasingly limited water supply—are very likely to be further stressed by rising temperatures.

Strategic Actions

In addition to continuing to implement the ongoing climate programs described in the introduction to this chapter, EPA intends to undertake the following efforts in the Northwest Region:

- **Sustainability**
  - Encourage sustainable infrastructure approaches.

    - Implement water conservation measures.
    - Expand use of GI.
    - Encourage communities and utilities to conduct vulnerability assessments and implement resulting recommendations.
- **Water Quality**
    - Implement water quality programs factoring in climate change to reduce stress on the ecosystem.
- **Collaboration**
    - Collaborate with the LCCs.
    - Engage in Western Governors Association climate adaptation activities.
    - Partner with the CSC and other federal agencies.
    - Engage tribes in federal climate conversations and activities.

Montane Region

The Montane region, in EPA Regions 8, 9, and 10, includes three glaciated mountain ranges: the Rocky Mountains, Sierra Nevada, and the Cascades. These areas are unique in that they rely on winter snow accumulation for their water supply. Sensitive ecological communities include bogs and fens. Montane glaciers and snowfields are reservoirs of water for the human populations and ecological communities at lower elevations.

Most ecosystems in the North American Montane Region are predicted to slowly migrate and shift their distribution toward the north in response to warming temperatures. However, the alpine areas are often distributed as small, isolated regions surrounded by other habitats. These areas can be disconnected from each other by wide stretches of land used for timber production, ranching, or other uses. Instead of shifts in latitude, alpine vegetation and animals will be limited to shifts in altitude, unless connections between suitable habitats can be made. [Jackson, 2006]

Region 8: http://www.epa.gov/region8/climatechange/
Region 9: http://www.epa.gov/region09/climatechange/
Region 10: http://yosemite.epa.gov/r10/water.nsf/homepage/water

Goals

The goal of EPA in the Montane Region is to protect the water quality and biological integrity of the Montane Region and increase the region's resilience to climate change, through water quality and habitat protection and restoration.

Strategic Issues

- A warmer climate will cause lower elevation habitats to move into higher zones, encroaching on alpine and sub-alpine habitats.
- High-elevation plants and animals will lose habitat area as they move higher, with some "disappearing off the tops of mountains."
- Rising temperatures will increase the importance of connections between mountain areas.
- Rising temperatures may cause mountain snow to melt earlier and faster in spring, shifting the timing and distribution of runoff. This in turn affects the availability of fresh water for natural systems and for human uses. Earlier melting leads to drier conditions for the balance of the water year, with increased fire frequency and intensity.
- Water supplies will become increasingly scarce, calling for tradeoffs among competing uses and leading to conflict.
- Increased frequency and altered timing of flooding will increase risks to people, ecosystems, and infrastructure.
- Projected increases in temperature, evaporation, and drought frequency add to concerns about the region's declining water resources.
- Climate change is likely to affect native plant and animal species by altering key habitats such as the wetland ecosystems known as montane fens or playa lakes.

Strategic Actions

In addition to continuing efforts in core climate change programs described in the introduction to this chapter, specific actions relative to the Montane Region include:

- Increase protection and restoration of wetlands to optimize percolation of surface water into ground water.

- Increase protection and restoration of riparian areas to reduce erosion during storm events and snow melt periods and thereby protect water quality.
- Increase protection of headwater streams and wetlands to protect the quality of montane water sources in the midst of precipitation and runoff-timing uncertainties.
- Collaborate with the USFWS, other DOI agencies, states, tribes, and others involved in LCCs in efforts to develop landscape-scale strategies to address climate change issues on a bio-regional basis.
- Coordinate climate change adaptation actions with federal agencies (given the large amount of federal agency holdings in the Montane Region), landholders, and others.
- Partner with other federal agencies to coordinate and leverage climate research and other activities.
- Engage tribes in federal climate conversations and activities.

Alaska Region

Over the past 50 years, Alaska has warmed at more than twice the average rate of the rest of the United States. Its annual average temperature has increased 3.4°F, while winters have warmed by 6.3°F. The higher temperatures are already causing earlier spring snowmelt, reduced sea ice, widespread glacier retreat, and permafrost warming. The observed changes are consistent with climate model projections of greater warming over Alaska, especially in winter, as compared to the rest of the country. Climate models also project increases in precipitation over Alaska. Simultaneous increases in evaporation due to higher air temperatures, however, are expected to lead to drier conditions overall, with reduced soil moisture. Average annual temperatures are projected to rise between 5 and 13°F by late this century. Increasing acidification of Alaskan waters presents a clear threat to Alaska's commercial fisheries and subsistence communities (USGCRP, 2009a).

Goals

- Design and build infrastructure that can withstand warmer conditions and thawing permafrost, flooding, and fire.
- Ensure adequate water supplies for communities dependent on disappearing sources.

- Protect water quality to reduce stress on the ecosystems.

Region 10: http://yosemite.epa.gov/r10/water.nsf/homepage/water

Strategic Issues

- Longer summers and higher temperatures are causing drier conditions, despite trends in increased precipitation. Insect outbreaks and wildfires are increasing with warming.
- A warmer climate will cause freshwater and saltwater species to move further north or into higher zones.
- As permafrost continues to thaw and temperatures rise, some lakes and ponds are beginning to disappear. This impacts drinking water sources and reduces wetland habitat while presenting a challenge for the ecosystem and the people who depend on its natural resources.
- Permafrost thaw has also caused numerous land slumps along riverbanks, which can have an impact on water quality (increasing turbidity) with documented impacts to drinking water in some Alaskan communities.
- Coastal storms increase risks to villages and fishing fleets. The combination of losing their protective sea ice buffer, increasing storm activity, and thawing coastal permafrost is causing some coastal communities to crumble into the sea. Increasing storm activity delays barge operations that supply coastal communities with fuel. The increased storm intensity puts fishing fleets at higher risk.
- Displacement of marine species will affect key fisheries. Thawing sea ice is moving the location and limiting the extent of plankton blooms. As plankton moves to deeper waters, it is less available to species and the food chain that depends on it, including humans dependent on these species for subsistence or economic livelihood.
- Thawing permafrost damages roads, runways, water and sewer systems, and other infrastructure.
- Opening of the Arctic from melting sea ice will create new opportunities for shipping, resource exploration and extraction, and tourism; there may be challenges caused by the increased traffic. Other key issues are the potential for the introduction of invasive species, impacts on subsistence activities, and national security concerns.

Strategic Actions

In addition to continuing to implement the ongoing climate programs described in the introduction to this chapter, EPA intends to undertake the following work in Alaska:

- **Infrastructure**
  - Encourage sustainable infrastructure approaches.
  - Encourage communities and utilities to conduct vulnerability assessments and implement resulting recommendations.
  - Encourage energy-efficient motors and pumps in infrastructure to reduce GHG emissions.
- **Water Quality and Water Supply**
  - Expand use of GI to delay stormwater runoff, mimic timing closer to the natural regime, and increase infiltration.
- **Collaboration**
  - Collaborate with the Alaska Climate Change Executive Roundtable and the LCCs and Climate Science Center in Alaska.
  - Partner with other federal agencies to coordinate and leverage climate research and other activities.
  - Engage tribes in federal climate conversations and activities.
  - Work with key federal, state, local, and tribal governments to assist communities that are evaluating relocation options as potential adaptation actions.

Caribbean Islands Region

Puerto Rico and the U.S. Virgin Islands, part of EPA Region 2, are located in the northeastern Caribbean Sea and are of volcanic origin. Puerto Rico (PR), including its offshore islands, covers a total area of 3,435 square miles. The main island of PR has three principal physiographic areas: the alluvial coastal plains, karst, and the central mountainous interior. Land surface elevations range from mean sea level to 4,389 feet above mean sea level. PR is home to approximately 3.9 million people, 70% of whom reside in coastal areas. Annual rainfall in PR ranges from about 30 inches in the western end of the south coast to about 160 inches near the top of the El Yunque Rainforest. Surface water provides approximately 75% of the population's freshwater needs. However, aquifers also play an important role in providing fresh water, especially to populations in the south coast and to the industrial sector.

Region 2: http://www.epa.gov/region2/climate/

The U.S. Virgin Islands (USVI), including the islands of St. John, St. Thomas, and St. Croix, cover a total area of 133 square miles. St. Thomas and St. John are characterized by steep topography while St. Croix is characterized by lower hills. Precipitation is the only natural source of fresh water on the islands. The population of the USVI relies on rooftop-rainfall catchments, large-scale desalination of seawater, and ground water.

The sensitive coastal ecosystems and critical infrastructure of the Caribbean Islands face difficulty due to sea level rise, tropical storms, and flooding from heavy rain. Coral reefs are under stress from warmer temperatures and ocean acidification. Water supplies are threatened due to both drought and saline contamination of aquifers.

Goals

The long-term goals of EPA in the Caribbean Region include:

- Work with partners to understand the vulnerability of coastal wetlands and their migration potential, and to protect the most vulnerable areas.
- Work with partners to understand the vulnerability of coastal communities and waterrelated infrastructure and to increase their resilience to extreme storm events.
- Increase understanding of the role of multiple stressors plus climate change on ecosystems and water-related infrastructure.

Strategic Issues

- Areas with limited ability for wetlands migration will see marked reductions in their ability to provide ecosystem services and will be increasingly vulnerable to intense storm damage in the future.
- Critical infrastructure (e.g., ports, airports, power plants, and sewage treatment facilities) in PR and the USVI located in the coastal zone will be vulnerable to storm surges, sea level rise, and the simultaneous occurrence of both.
- Many hurricanes and coastal inundations are accompanied by heavy rains and river/stream floods, which impact water quality and stream morphology.

- Rising sea levels cause intrusion of salt water into the underground freshwater lens, contaminating the supply of usable ground water and reducing the freshwater supply for the Caribbean Islands.
- Higher ambient water temperatures and degradation of water quality, including adjustments in pH due to acidification, may affect production rates of aquaculture facilities and their susceptibility to diseases such as microbial infections and parasitic infestations. Presently, there are no aquaculture facilities operating in the Caribbean, although a few NPDES permits have been issued.
- Recent events of increased sea surface temperatures have caused stress to coral reefs in the USVI and PR. Increasing sea surface temperatures have the potential to reduce the stability of corals, especially in the presence of stresses from the existing land-based sources of pollution.
- Ocean acidification may potentially diminish the quality of the reefs by impeding the calcification process, increasing carbon in the water, altering ocean chemistry, and making calcium less available for calcification. Lower pH could also accelerate erosion of existing structures.
- Longer periods of drought are expected to occur and may produce an increase in the energy and costs associated with the production of drinking water. This will be particularly pertinent in the USVI, where desalination is one of the main sources of drinking water.

Strategic Actions

In addition to continuing to implement the ongoing climate programs described in the introduction to this chapter:

- Support the development of LIDAR images for the Caribbean Region in order to provide more refined data for modeling purposes.
- Support the integration of climate change considerations into FEMA, Commonwealth, and municipal hazard mitigation plans.
- Promote increased use of GI in the Caribbean to enhance resilience by absorbing and infiltrating stormwater and preventing flooding and pollution impacts by providing outreach and education to the public and to design and building professionals. Support PR in the development of tsunami-ready communities.
- Partner with the Caribbean Coastal Ocean Observing System (CariCOOS), the National Weather Service, and other relevant federal

and Commonwealth agencies to disseminate information and provide outreach to managers of PR's water infrastructure with regard to current trends.

- Support and encourage increased resilience of water infrastructure through physical upgrades, geomorphic feature protection (e.g., barrier islands, mangrove islets, eolianites, beach rock, and dunes), building code revisions, and working with insurance companies so they implement disaster risk reduction measures in the underwriting criteria of their policies.
- Engage environmental justice (EJ) populations in education on climate change impacts and planning for climate change adaptation.

Pacific Islands Region

The Pacific Islands region in EPA Region 9 encompasses the Hawaiian Islands as well as the U.S.-affiliated Pacific Islands, including the territories of American Samoa, the Commonwealth of the Northern Mariana Islands (CNMI), and Guam.

The Pacific Islands are more vulnerable to climate change than nearly any other region in the United States. Key vulnerabilities include availability of fresh water, adverse impacts to coastal and marine ecosystems, and exposure to hazards, including sea level rise and inundation.

Region 9: http://www.epa.gov/region09/climatechange/

Goals

- Design and build infrastructure that can withstand storms, flooding, salt spray, and fire.
- Protect existing drinking water supplies and ensure adequate supplies for communities dependent on disappearing sources.
- Encourage communities and utilities to conduct vulnerability assessments and implement resulting recommendations.
- Work with local governments on disaster planning and response, and long-term plans to protect infrastructure and human safety.
- Protect coral reefs, mangroves, and other sensitive ecosystems.

- Educate local and cultural leaders on the impacts of climate change and engage them in planning for climate change adaptation.

Strategic Issues

- Rising sea levels, higher sea temperatures, and ocean acidification associated with climate change are further degrading coral reefs already stressed by overfishing and pollution. Their loss diminishes ecological heritage, shoreline protection, and food supply from the sea, and results in a decline in income from ecotourism in the Pacific Island communities, where tourism is one of the largest industries.
- The western Pacific already experiences the highest rate of Category 4 and 5 storms. Climate change may bring more frequent and higher energy storms resulting in potentially catastrophic damage to island infrastructure. This degree of damage could cripple the economies of Pacific Island communities for significant periods of time, not only impairing economic development, but also the ability of local governments to ensure delivery of basic water and sewer and other public health services.
- **Sea level rise has multiple implications for Pacific Island communities:**
  - For the low-lying atolls, entire islands may be submerged within a generation and may result in environmental refugees seeking new homes.
  - For some low-lying islands, sea level rise can result in "wash over," in which islands, or portions of islands, are submerged by waves during large storm events. This results in saltwater contamination of agricultural lands, significantly decreasing the productivity of those lands. This loss of agricultural productivity has an acute impact on the largely subsistence-based economies of these communities.
  - For many of the islands, sea level rise has an immediate and accelerated impact on coastal erosion, which affects water quality, coral reef health, coastal infrastructure, available land, and culturally significant sites.
  - Sea level rise increases the potential for saltwater intrusion into the sole source aquifers upon which many Pacific Islands rely for drinking water. There are few or no readily accessible alternative

drinking water options when a community is confronted with the loss of productivity of a sole source aquifer.

Strategic Actions

In addition to continuing to implement the ongoing climate programs described in the introduction to this chapter:

- Work with local, state, and federal agencies, as well as local educational institutions, to ensure protocols are in place to identify key drinking water resources, monitor water quality, and develop long-term drinking water protection and management plans.
- Work with local, state, and federal agencies to leverage capital improvement funds to develop water and wastewater infrastructure designed to be resilient to the effects of climate change.
- Develop biological criteria as a component of water quality standards as a tool for coral reef protection.
- Use permitting authorities and enforcement to protect drinking water and near-shore water quality consistent with the requirements of the CWA.
- Work with local, state, and federal agencies to invest in local utility managers and employees for the long term so they have the skills and resources to consistently protect public health and safety, even in the event of catastrophic storm events. Collaboratively identify best management practices that are institutionalized through standard operating procedures.
- Work with local, state, and federal agencies to reduce reliance on fossil fuels through energy audits, conservation incentives, and investment in renewable energy sources. This approach will reduce water quality impacts associated with oil spills and develop an energy infrastructure that may be more resilient to severe storm events.
- Work with local, state, and federal agencies, as well as local educational institutions, to reduce stressors to coral reef health (e.g., sedimentation and impacts from fishing and recreation) and to protect coral reef ecosystems in perpetuity.
- Work with local, state, and federal agencies to build awareness of the potential effects of climate change and opportunities to reduce GHG emissions and adapt to impacts.
- Engage cultural leaders and EJ populations in education on climate change impacts and planning for climate change adaptation.

VI. Cross-Cutting Program Support

A. Goal 17: Communication, Collaboration, and Training

Strategic Action 51: Continue building the communication, collaboration, and training mechanisms needed to effectively increase adaptive capacity at the federal, tribal, state, and local levels.

The NWP intends to continue building the communication, collaboration, and training mechanisms needed to effectively increase adaptive capacity at the federal, tribal, state, and local levels, including nongovernmental and private sector stakeholders.

Communicating Effectively

Communication involves three elements: the audience, the message, and the medium. This *2012 Strategy* describes the messages, including:

- Climate change poses threats to water resources and the NWP's mission.
- Ecosystem services associated with water are valuable resources for modulating climate impacts.
- Water management strategies can reduce greenhouse gas emissions and increase resilience to climate change.
- Programmatic actions are being taken to address climate change.
- Information and tools are needed to inform action.
- Collaboration is essential for shared learning and problem solving.

Examples of Collaborative Partnership Programs

Infrastructure:

- Source Water Collaborative
- Rural Community Assistance Partnership
- Capacity Development Program
- Effective Utility Management

Ocean and Coastal Waters:

- National Ocean Council
- National Estuary Programs

Watersheds and Wetlands:

- Healthy Watersheds Initiative

Water Quality:

- Green Infrastructure Initiative

As described earlier, there are many stakeholders with interests and responsibilities for protecting the nation's water resources. Some of the audiences the NWP intends to communicate with include:

- State and tribal co-regulators who need information and tools to adapt their programs.
- The water utility operators who need tools to calibrate their design and management practices for protecting infrastructure from climate change impacts.
- Natural resource professionals who protect water quality and ecological integrity from compounding stressors, including climate change.
- Tribal communities that have geographically and culturally specific challenges for protecting and preserving their freshwater resources and communities.
- Economically disadvantaged communities that may already have a deficit in the ability to respond to impacts.
- Communities that are at risk from sea level rise, flood, and drought.
- The public and stakeholders who want to know how the federal government is addressing climate change.
- The private sector who are working to protect their investments while responsibly managing natural resources.
- Federal agencies with which EPA collaborates.

EXAMPLES OF REGIONAL COLLABORATION GOALS

- **Region 1 Federal Partners Group**
  http://www.epa.gov/region1/eco/energyefforts-epane.html.
- **Region 4 Southeast Natural Resources Leadership Group**
  http://www.epa.gov/region4/topics/envmanagement/senrlg/index.htm.
- **Region 9 Water-Energy Team of the California Climate Action Team (WET-CAT)**
  http://www.climatechange.ca.gov/wetcat/index.html.
- **Great Lakes Statement of Common Purpose**
  http://collaborate.csc.noaa.gov/nroc/default.aspx.
- **Gulf of Mexico Alliance Action Plan II for Healthy and Resilient Coasts, 2009-2014**
  http://www.gulfofmexicoalliance.org/pdfs/ap2_final2.pdf?#Page=8.

In addition to the avenues discussed throughout this document for training and tool development, the NWP intends to provide communication outlets including:

- National Water Program Climate Change & Water Website
- *EPA Climate Change & Water News* E-Newsletter
- Climate Ready Water Utilities website and toolbox
- Webcasts and Webinars – to provide opportunities for targeted training
- Ongoing dialogue forums with stakeholder groups and co-regulators such as the State-Tribal Climate Change Council (STC3)

EXAMPLES OF FEDERAL COLLABORATIVE FORUMS

- Interagency Climate Change Adaptation Task Force, Freshwater Workgroup
- White House Office of Science & Technology Policy Subcommittee on Water Availability and Quality
- U.S. Global Change Research Program
- Climate Change Adaptation Work Group

- Speaker Series – for EPA employees to hear from experts

- Annual reports and periodic updates

The NWP intends to continue working with partners and stakeholders to develop communication mechanisms to expand access to information and resources for general and targeted audiences.

Working in Collaboration

Existing EPA partnership programs provide ready access to networks of key entities and can be leveraged to address the challenges posed by climate change while minimizing the "overhead" involved in collaboration. Examples of existing programs and initiatives are referenced throughout this document, and some are noted in the text box below.

FEDERAL SOURCES OF CLIMATE CHANGE INFORMATION

EPA

Water & Climate Change: http://water
Climate Ready Water Utilities: http://www.water.epa.gov/crwu
Climate Ready Estuaries: http://www.epa.gov/cre/
Climate Change: http://www.epa.gov/climatechange/

NOAA

Climate Service: http://www.climate
RISAs: http://www.research.noaa.gov/climate
Coasts: http://www.csc.noaa.gov/digitalcoast/tools/index.html

Interagency

USGCRP: http://www.globalchange.gov
Smartcoasts: http://stormsmartcoasts.org/
http://www.epa.gov/adr/index.html

The NWP intends to work to expand opportunities for dialogue through both formal and informal discussion. For example, ACWI and the National Drinking Water Advisory Committee (NDWAC) are key Federal Advisory

Committees (FACAs). The NWP's STC3 is an important discussion forum with state and tribal co-regulators. Collaboration with sector partnerships is a particularly important avenue for promoting research, pilots, and communication, including associations such as the Water Utility Climate Alliance (WUCA), WRF, and the WERF.

Climate impacts are local, as are adaptation strategies; hence, many EPA Regions are building collaborations with state, tribal, and local government agencies as well as with other federal agencies to more effectively deliver services. Information on Regional partnerships can be found in Chapter V, *Geographic Climate Regions*.

Federal partnerships are also essential to leveraging resources and building national capability for adaptation. The NWP intends to continue to strengthen and expand our coordination on climate change adaptation and mitigation with other federal agencies at both the national and regional levels.

Delivering Tools and Training

Many of the Strategic Actions throughout this *2012 Strategy* are driven by the over-riding need to improve the translation of climate impact projections into materials tailored for NWP partners and constituents, including regionally specific information. The NWP intends to work to make information available, including training to help practitioners apply new tools. The NWP intends to collaborate with various forums for delivering the information and training.

National Water Program Implementation

- The core CWA, SDWA, and other statutorily authorized programs within the NWP have training forums such as the NPDES Permit Writers training, the Watershed Academy, the Water Quality Standards Academy, and the Drinking Water Capacity Development Program that reach out to practitioners.
- Partner organizations host and co-sponsor training sessions, such as those based on the *Clean Water and Safe Drinking Water Infrastructure Sustainability Policy* to promote best practices for effective utility management, energy management, and advanced asset management, and related topics such as GI and LID.
- Conflict Resolution is a field that can help to build skills for collaborating and consensus building for working effectively with stakeholders. The NWP intends to work with partners and stakeholders to draw on the expertise and resources of the Conflict

Prevention and Resolution Center (CPRC) within EPA's Office of General Counsel as well as the Regional Alternative Dispute Resolution (ADR) Specialists to conduct training.

Decision Support

- The USGCRP delivers science and science translation to inform adaptation planning.
- Federal partners are building regional capabilities, such as NIDIS, NOAA RISAs, and the LCCs and CSCs launched by DOI.
- The interagency CCAWWG, under the leadership of the Bureau of Reclamation, is working to establish a training program for water resource managers.

B. Goal 18: Tracking Progress and Measuring Outcomes

Strategic Action 52: Adopt a phased approach to track programmatic progress towards Strategic Actions; achieve commitments reflected in the Agency *Strategic Plan*; work with an EPA workgroup to develop outcome measures.

Tracking and measuring progress towards a stated goal provides information about the efficacy of the actions taken to inform adaptive management; provides a way to share information and lessons learned with others working toward similar ends; and provides transparency to stakeholders who have an interest in the process or outcome. Devising meaningful and practical indicators for tracking progress, however, is complex. It is preferable to measure *outcomes* rather than *outputs*, but outcomes often take many years to realize and may be hard to quantify. The NWP intends to work to develop and refine these measures, including ways to measure outcomes. Tracking progress for climate change adaptation poses its own challenges, including how to evaluate avoided losses.

Table 5. Phases of Adaptive Management

NWP Phases	Explanation	Examples of Evidence of Achievement	NWP Status
1. Initiation	Conduct a screening assessment of potential implications of climate change to mission, programs, and operations.	• Preliminary information is developed to evaluate relevance of climate change to the mission or program; a decision is made as to whether to prepare a response to climate change; further exploration of climate change implications has been authorized. • Accountabilities and responsibilities are assigned at appropriate levels within the organization and resources are available to develop a more in-depth assessment.	1
2. Assessment	Conduct a broader review to understand how climate change affects the resources in question. Work with stakeholders to develop an understanding of the implications of climate change to the mission, programs, and operations.	• Review science literature and assessments to understand how climate change affects the resources being protected (threat to mission); Engage internal staff and external stakeholders in evaluation. • Identify climate change issues and concerns and communicate with internal and external stakeholders and partners. • Identify which specific programs are threatened and what specific information or tools need to be developed. • Communicate findings to partners and stakeholders and engage them in dialogue on building adaptive capacity.	2
3. Response Development	Identify changes necessary to continue to reach program mission and goals. Develop initial action plan. Identify and seek the research, information, and tools needed to support actions. Begin to build the body of tools, information, and partnerships needed to build capacity internally and externally.	• Develop initial program vision and goals for responding to climate change. • Identify needed response actions or changes that will allow the organization to begin to address climate impacts on its mission. • Initiate strategies and actions in a few key areas to begin to build organizational ability to use climate information in decision processes. • Identify program partners' needs for building adaptive capacity. • Begin working with an external "community of practice" to engage in tool and program development. • Rudimentary methods are put in place to track progress. • Develop a research strategy and partnerships to obtain additional needed research.	3

NWP Phases	Explanation	Examples of Evidence of Achievement	NWP Status
4. Initial Implementation	Initiate actions in selected priority programs or projects.	• Make it clear within the organization that incorporating climate change into programs is critical. • Initiate actions and plans identified in Step 3. • Initiate cooperative projects with partners. • Develop a range of needed information and tools. • Begin to institute changes to incorporate climate change into core programs. • Some program partners have begun to implement response actions.	4
5. Robust Implementation	Programs are underway and lessons learned are being applied to additional programs and projects.	• Lessons learned are evaluated and strategies are refined. • Efforts are initiated to consider climate change in additional, or more complex, program elements. • Continue to institute institutional changes to incorporate climate change into core programs. • External communities of practice are in place to support ongoing capacity development.	5
6. Main-streaming	Climate is an embedded, component of the program.	• The organization's culture and policies are aligned with responding to climate change. • All staff have a basic understanding of climate change causes and impacts. • All relevant programs, activities, and decision processes intrinsically incorporate climate change. • Methods for evaluating outcomes are in place.	6
7. Monitoring and Adaptive Management	Continue to monitor and integrate performance, new information, and lessons learned into programs and plans.	• Progress is evaluated and needed changes are implemented. • As impacts of climate change unfold, climate change impacts and organizational responses are reassessed.	7

NWP Phased Approach for Indicators of Progress

Currently, the most amenable approach for evaluating progress is to assess institutional progress toward becoming a resilient and adaptive program. The NWP is adopting a phased approach that uses indicators of progress and emphasizes peer-to-peer learning rather than a top-down mandate. A similar approach is in use in the United Kingdom (UK DEFRA, 2010).

Initially, the NWP Phased Approach intends to track the NWP's institutional *process* and *progress* in incorporating climate change considerations into EPA programs. *Outputs* will not be counted per se; rather, the collectivity of actions and their products will demonstrate *the weight of evidence* for determining the status of adaptation activities. An annual reporting process will assemble information for evaluating and publicly reporting progress. The elements to be assessed include progress toward achieving the stated Goals and Strategic Actions (Headquarters programs) and progress toward implementing Regional strategies. The NWP intends to work with its State-Tribal Climate Change Council and other partners to refine this approach and develop a model that could be useable by others at their discretion.

Table 5 presents a summary of the seven phases. Recognizing that it may take years or decades to achieve adaptive preparedness and resilience, the NWP designed phases for which progress could be demonstrated within a relatively short time frame (1 to 3 years).

In addition to the process to track progress described in Table 5, EPA's *2011–2015 Strategic Plan* includes measures for climate change adaptation and mitigation actions, listed in Table 6. This *2012 Strategy* reflects the NWP's commitment to achieve these measures. An EPA workgroup has undertaken a process to refine and update the Agency's measures to reflect outcomes toward desired objectives. The NWP intends to work with the above mentioned EPA workgroup to develop outcome measures applicable to the NWP.

C. Goal 19: Climate Change and Water Research Needs

Strategic Action 53: Work with the EPA's Office of Research and Development, other water science agencies, and the water research community to further define needs and develop research opportunities to deliver the information needed to support implementation of the *2012*

***Strategy*, including providing the decision support tools needed by water resource managers.**

Table 6. *2011–2015 EPA Strategic Plan* National Water Program Commitments

<table>
<tr><td colspan="2">Goal 1: Taking Action on Climate Change and Improving Air Quality. Reduce greenhouse gas emissions and develop adaptation strategies to address climate change, and protect and improve air quality.</td></tr>
<tr><td colspan="2">Objective 1.1: Address Climate Change. Reduce the threats posed by climate change by reducing greenhouse gas emissions and taking actions that help communities and ecosystems become more resilient to the effects of climate change.</td></tr>
<tr><td rowspan="4">Water-Related Strategic Measures:</td><td>• By 2015, additional programs from across EPA will promote practices to help Americans save energy and conserve resources, leading to expected greenhouse gas emissions reductions of 740.1 MMT CO_2 Eq. from a baseline without adoption of efficient practices.
The WaterSense Program will contribute to achieving greenhouse gas reduction goals through 2015.</td></tr>
<tr><td>• By 2015, EPA will integrate climate change science trend and scenario information into five major scientific models and/or decision-support tools used in implementing Agency environmental management programs to further EPA's mission, consistent with existing authorities.
Under the CRWU initiative, the NWP will deploy an upgraded version of the CREAT, as well as a comprehensive toolbox of water-related climate resources by the end of 2012, to better assist water and wastewater utilities in becoming more resilient to climate change.</td></tr>
<tr><td>• By 2015, EPA will account for climate change by integrating climate change science trend and scenario information into five rule-making processes to further EPA's mission, consistent with existing authorities.
The NWP will incorporate climate change considerations in the development and implementation of a rulemaking by 2015.</td></tr>
<tr><td>• By 2015, EPA will build resilience to climate change by integrating considerations of climate change impacts and adaptive measures into five major grant, loan, contract, or technical assistance programs to further EPA's mission, consistent with existing authorities.
The NWP will help NEP grantees consider as a potential priority climate adaptation and resilience in their Comprehensive Conservation and Management Plans s and develop climate adaptation plans and implementation strategies where considered a priority.</td></tr>
</table>

This section describes the types of research questions that need to be addressed to support the Strategic Actions in this *2012 Strategy*. In general, research for adaptation should provide decision support to manage risk in an evolving context under ranges of uncertainty. Implementation of this strategy will incorporate new research and tools as they become available.

The NWP collaborates with and relies on the broader research community, including EPA ORD, federal science agencies (e.g., USGS, NOAA, USGCRP), drinking water and water quality research associations (e.g., WRF, WERF), academia, and others. The NWP is also a member of CCAWWG, a "working level" forum for sharing expertise and planning to build climate adaptation tools and methods across federal agencies. These collaborations have already produced a range of reports and inventories on research needs and activities. (See: CCAWWG, 2011; WRF, 2011; EPA-ORD, 2012.) The NWP intends to continue to work with the water research community to further define needs and develop collaborative and coordinated research opportunities to deliver the information needed by water resource managers.

Cross-Cutting Research Needs

A. Data: Update data for precipitation, storm frequency, and streamflow, and develop new methods for analyzing projected changes, in collaboration with other federal agencies.

1. Of particular concern are the storm frequency, duration, and intensity estimates and low-flow conditions in rivers and streams at the HUC 12 watershed level.
2. Improve methods to address non-stationarity, particularly improving clarity of precipitation data used in wastewater, drinking water, and stormwater management systems design, operation, and planning (e.g., TP40, Atlas 14).
3. Enhance flow estimation using NHDPlus.

B. Decision Support: Integrate non-stationarity and recent data into decision support tools for water utilities and water quality managers to use in planning across a range of plausible climate change scenarios.

1. Research Translation: Produce annual or biennial synthesis reports of recent research and implications for decision-makers to inform the water resource management community.
2. Modify climate model outputs that can be used as inputs for hydrologic and management models at the spatial and temporal scales relevant to decision-makers.

3. Develop regionally specific information (include description of observed and projected impacts, scenarios, etc.) for communicating with communities and tribes.
4. Develop models that integrate hydrology, land cover, air quality, and economics for comprehensive assessment and comparison of climate change mitigation and adaptation policies for local, state, and federal governments.
5. Develop a rapid response protocol to incorporate the results of the ongoing monitoring data into permitting, planning, and resource allocation decisions.
6. Develop tools for prioritizing response actions that take into account potential for both adaptation and GHG mitigation, especially for wetlands protection and restoration.

C. **Metrics:** Develop measures and metrics to track and determine progress in climate change adaptation and preparedness.

Research to Support Infrastructure

A. **Water Demand Management:** Design metrics for water and energy efficiency in key sectors (e.g. municipal use; energy production and agriculture). Produce methods and technology transfers in various sectors to reduce water demand.

B. **Water Supply Management:** Develop alternative and nonconventional water supplies that will relieve pressure on freshwater sources and ensure the protection of current and future sources of drinking water.

C. **Energy-Water Nexus:** Develop zero-net energy strategies through life cycle analysis of water/energy consumption and optimization and co-generation.

D. **Aquifer Storage and Recharge:** Research into technologies to minimize mobilization of geologic chemicals/radionuclides and the formation of new drinking water contaminants by injectate that is already treated to national drinking water standards. Consider natural attenuation of microbes in different soil and geologic profiles and disinfectant byproducts from treated injectate. Also identify configurations that minimize adverse effects on surface water/ground water interchange (e.g., that maintain healthy instream flows to support aquatic habitats).

E. **Economics:** Conduct cost-benefit analysis of climate change adaptation strategies. Evaluate the cost of adapting versus the comparative costs of business-as-usual approaches. Calculate the value of built infrastructure at risk from climate change, especially from sea level rise and flooding, and use this information in economic assessments of potential adaptation strategies.

Research to Support Watersheds and Wetlands

A. **Monitoring:** Identify aquatic ecosystem responses to changes in temperature, precipitation, and sea level rise. Identify water chemistry changes including possible acidification effects that may be occurring in freshwater and estuarine systems. Develop water monitoring designs to track parameters relevant to climate change impacts. Identify and measure shifts over time in the condition of water resources attributed to climate change.
B. **Hydrology:** Improve the understanding of climate change on the hydrologic function of wetlands and in providing ecological services. Increase the understanding of the hydrological connections between surface water and ground water to inform IWRM. Model potential changes to flood regulation, ground water recharge, and surface water base flow, given scenarios of wetlands loss, including from increased ground water pumping. Assess different types of wetlands' capacity to adapt to climate change.
C. **Co-benefits:** Characterize co-benefits of healthy watersheds, GI, and site conditions where GI is cost-effective and where it is not:
    1. Identify climate change mitigation and adaptation strategies that lead to water quality improvements, such as increased ground water recharge and stormwater runoff mitigation and reduced cost for stormwater management and green space connectivity.
    2. Develop method to measure carbon sequestration potential for aquatic ecosystems (e.g., wetland types, and forested watersheds).

Research to Support Coastal and Ocean Planning

A. **Ocean Acidification:** Understand likely impacts of ocean acidification to coastal systems/system components, and identify and fill information gaps. Assess relative vulnerabilities in order to identify sites appropriate for action to increase coral reef resiliency.

B. **Sea Level Rise:** Accurate mapping of relative historic and projected sea level rise and its impacts. Determine which coastal wetlands and ecosystems to protect or restore and those that are "lost."
C. **Temperature:** Investigate potential impacts of climate change, such as warming water temperatures on eutrophication and ecology.

Research to Support Water Quality

A. **Pathogens:** Evaluate potential changes in exposure factors and assessment methods for waterborne pathogens that result from climate change. Develop models to better understand how increased water temperature affects pathogen survival and proliferation, drinking water treatment, and sanitary waste treatment requirements based on water quality based effluent limitations or effluent limitation guidelines. Identify contaminants that may more greatly affect public water system noncompliance by increases or decreases in precipitation or ground water levels.
B. **Precipitation:** Identify impacts from changes such as extreme precipitation events that may increase sediment loading or scouring, nutrient, pathogen, and toxic contaminant loads to water bodies.
C. **Nitrogen Cycle:** Assess air-water interactions (i.e., sources and sinks) of nitrogen and develop strategies to reduce impacts to aquatic ecosystems and ground and surface drinking water sources.
D. **Flow:** Characterize ecological flow criteria for aquatic species to protect designated uses, given climate change intensifying the competition for finite water resources. The criteria may be useful in developing TMDLs.

Research to Support Tribes

A. **Traditional Ecological Knowledge:** Strengthen the ability to incorporate tribal traditional knowledge into adaptation strategies relevant to tribes.
B. **Overall:** Include development of tribal-specific elements in overall research strategies to understand climate change impacts and to develop adaptation strategies.

VII. APPENDICES

Appendix A: Principles for an Energy Water Future – The Foundation for a Sustainable America

Principles for an Energy Water Future

A Foundation for a Sustainable America

The nexus between energy and water is an increasingly important area for focus. There are significant societal and environmental benefits to be derived from improving coordination between the two sectors. Government should take a leadership role in this relationship and lead by example. EPA is proposing principles for government, service providers, and ratepayers to foster valuable collaboration in both the water and energy sectors to work together to meet our water and energy needs nationally and locally. The principles also serve as a reminder that rising water treatment costs or necessary tradeoffs such as stricter water treatment levels can be mitigated by efforts elsewhere such as reducing demand for energy and water.

Efficiency in the use of energy and water should form the foundation of how we develop, distribute, recover, and use energy and water. EPA supports:

- Encouraging energy and water efficiency by the ratepayer through the use of efficient products, like ENERGY STAR and WaterSense labeled products, supplemented by informed and wise use of resources.
- Improving system-level energy and water efficiency by water, wastewater, stormwater, and energy utilities and encouraging strategic investments in efficiency.
- Using full-cost rate structures while ensuring access to clean and safe water for low income households.
- Recognizing and reducing the embedded water and energy in manufactured and agricultural products.
- Relying on education and outreach, in collaboration with local communities, to be at the forefront of encouraging efficiency.

The exploration, production, transmission, and use of energy should have the smallest impact on water resources as possible, in terms of water quality and water quantity. EPA supports:

- Reducing consumption or use of water for producing energy and fuels: reduce, recover, reuse, and recycle.
- Analyzing, recognizing, and minimizing any impacts on groundwater, water quality, water quantity, and the aquatic environment, including wetlands, when choosing between sources of energy.
- Practicing good stewardship to minimize potential impacts and avoid contaminants that reduce water's value or require additional energy for treatment.

The pumping, treating, distribution, use, collection, reuse, and ultimate disposal of water should have the smallest impact on energy resources as possible. EPA supports:

- Creating an energy efficiency management plan using established energy auditing tools.
- Establishing plans to repair leaks in water distribution and wastewater collection systems.
- Using nearby water sources where available, including rain harvesting and recycled water.
- Treating water to a level that matches the end use.
- Avoiding unnecessary transport of water and wastewater for treatment or disposal.

Wastewater treatment facilities, which treat human and animal waste, should be viewed as renewable resource recovery facilities that produce clean water, recover energy, and generate nutrients. EPA supports:

- Using wastewater and associated organic solids and treatment byproducts, such as methane gas, as a source of renewable energy that can be used by treatment plants to reduce net 'on-grid' energy use or to become zero net energy consumers.
- Using wastewater for irrigation, accounting for the nutrients in the water as a way to reduce the need for additional fertilizers.

- Recycling or reusing water for appropriate uses with no resulting loss of downstream use and habitat, minimizing energy used for treatment, and becoming a reliable source for the future.
- Extracting and recycling nutrients from wastewater.

The water and energy sectors – governments, utilities, manufacturers, and consumers – should move toward integrated energy and water management from source, production, and generation to end user. EPA supports:

- Encouraging the water and energy sectors – both governments and utilities – to continue to align themselves to breakdown institutional barriers, improve transparency, and maximize efficiencies.
- Encouraging government agencies to look across missions and private utilities to look across sectors to achieve integrated energy and water management, maximize efficiencies, and avoid unintended consequences.
- Encouraging partnerships between government and service providers to leverage and expand upon existing successes and institutions.
- Promoting transparency and collaboration related to research, funding, and policy within institutions and across sectors, which are essential and will help to leverage lessons learned and expand successes.

Maximize comprehensive, societal benefits. EPA supports:

- Articulating and recognizing the benefits for the larger sphere of influence of public and private investment – beyond direct cost savings – in energy and water efficiencies.
- Enhancing, promoting, and targeting financial incentives and other societal benefits, including market-based benefits such as rebates and government programs such as state revolving funds, taxes, and tax credits.
- Planning to build resiliency for climate change impacts on water infrastructure and water quality to minimize vulnerabilities.

Appendix B: Goals and Strategic Actions: Lead Offices[6]

Goals and Strategic Actions		Lead Office (& Partners)
Infrastructure: In the face of a changing climate, resilient and adaptable drinking water, wastewater, and stormwater utilities (i.e., the water utility sector) ensure clean and safe water to protect the nation's public health and environment by making smart investment decisions to improve the sustainability of their infrastructure and operations and the communities they serve, while reducing greenhouse gas emissions through greater energy efficiency.		
Goal 1: Build the body of information and tools needed to incorporate climate change into planning and decision making.	**SA1:** Improve access to vetted climate and hydrological science, modeling, and assessment tools through the Climate Ready Water Utilities program.	OGWDW (OWM)
	SA2: Assist wastewater and water utilities to reduce greenhouse gas emissions and increase long-term sustainability with a combination of energy efficiency, co-generation, and increased use of renewable energy resources	OWM (OGWDW)
	SA3: Work with the states and public water systems, particularly small water systems, to identify and plan for climate change challenges to drinking water safety and to assist in meeting health based drinking water standards.	OGWDW
	SA4: Promote sustainable design approaches to provide for the long-term sustainability of infrastructure and operations.	OWM (OGWDW)
Goal 2: Support Integrated Water Resources Management to sustainably manage water resources.	**SA5:** Understand and promote through technical assistance the use of water supply management strategies.	OWM (OGWDW)
	SA6: Evaluate and provide technical assistance on the use of water demand management strategies.	OWM (OGWDW)
	SA7: Increase cross-sector knowledge of water supply climate challenges and develop watershed specific information to inform decision making.	OW IO (All OW Offices)
Watersheds & Wetlands: Watersheds are protected, maintained, and restored to provide climate resilience and to preserve the ecological, social, and economic benefits they provide; and the nation's wetlands are maintained and improved using integrated approaches that recognize their inherent value as well as their role in reducing the impacts of climate change.		
Goal 3: Identify, protect, and maintain a network of healthy watersheds and	**SA8:** Develop a national framework and support efforts to protect remaining healthy watersheds and aquatic ecosystems.	OWOW
	SA9: Collaborate with partners on terrestrial ecosystems and hydrology so that effects on water quality and aquatic ecosystems are considered.	OWOW

(Continued)

Goals and Strategic Actions		Lead Office (& Partners)
supportive habitat corridor networks.	**SA10:** Integrate protection of healthy watersheds throughout the NWP core programs.	OWOW
	SA11: Increase public awareness of the role and importance of healthy watersheds in reducing the impacts of climate change.	OWOW
Goal 4: Incorporate climate resilience into watershed restoration and floodplain management.	**SA12:** Consider a means of accounting for climate change in EPA funded and other watershed restoration projects.	OWOW
	SA13: Work with federal, state, interstate, tribal, and local partners to protect and restore the natural resources and functions of riverine and coastal floodplains as a means of building resiliency and protecting water quality.	OWOW
Goal 5: Watershed protection practices incorporate Source Water Protection to protect drinking water supplies.	**SA14:** Encourage states to update their source water delineations, assessments or protection plans to address anticipated climate change impacts.	OGWDW
	SA15: Continue to support collaborative efforts to increase state and local awareness of source water protection needs and opportunities, and encourage inclusion of source water protection areas in local climate change adaptation initiatives.	OGWDW
Goal 6: EPA incorporates climate change considerations into its wetlands programs, including the CWA 404 program, as appropriate.	**SA16:** Consider the effects of climate change, as appropriate, when making significant degradation determinations in the CWA Section 404 wetlands permitting and enforcement program.	OWOW
	SA17: Evaluate, in conjunction with the U.S. Army Corps of Engineers, how wetland and stream compensation projects could be selected, designed, and sited to aid in reducing the effects of climate change.	OWOW
Goal 7: Improve baseline information on wetland extent, condition, and performance to inform	**SA18:** Expand wetland mapping by supporting wetland mapping coalitions and training on use of the new federal Wetland Mapping Standard.	OWOW
	SA19: Produce a statistically valid ecological condition assessment of the nation's wetlands.	OWOW
	SA20: Work with partners and stakeholders to develop information and tools to support long term planning and priority setting for wetland restoration projects.	OWOW

Goals and Strategic Actions		Lead Office (& Partners)
long term planning and priority setting that takes into account the potential added benefits for climate change adaptation and carbon sequestration.		
Coastal and Ocean Waters: Adverse effects of climate change along with collective stressors and unintended adverse consequences of responses to climate change have been successfully prevented or reduced in the ocean and coastal environment. Federal, tribal, state and local agencies, organizations, and institutions are working cooperatively; and information necessary to integrate climate change considerations into ocean and coastal management is produced, readily available, and used.		
Goal 8: Collaborate so that information and methodologies for ocean and coastal areas are collected, produced, analyzed, and easily available.	**SA21:** Collaborate so that synergy occurs, lessons learned are transferred, federal efforts effectively help local communities, and efforts are not duplicative or at cross-purposes.	OWOW
	SA22: Work within EPA and with the U.S. Global Change Research Program and other federal, tribal, and state agencies to collect, produce, analyze, and format knowledge and information needed to protect ocean and coastal areas and make it easily available.	OWOW
Goal 9: Support and build networks of local, tribal, state, regional and federal collaborators to take effective adaptation measures for coastal and ocean environments through EPA's geographically targeted programs.	**SA23:** Work with the NWP's larger geographic programs to incorporate climate change considerations, focusing on both the natural and built environments.	OWOW (Regions)
	SA24: Address climate change adaptation and build stakeholder capacity when implementing NEP Comprehensive Conservation and Management Plans and through the Climate Ready Estuaries Program.	OWOW
	SA25: Conduct outreach and education, and provide technical assistance to state and local watershed organizations and communities to build adaptive capacity in coastal areas outside the NEP and Large Aquatic Ecosystem programs.	OWOW

(Continued)

Goals and Strategic Actions		Lead Office (& Partners)
Goal 10: Address climate driven environmental changes in coastal areas and provide that mitigation and adaptation are conducted in an environmentally responsible manner.	**SA26:** Support coastal wastewater, stormwater, and drinking water infrastructure owners and operators in reducing climate risks and encourage adaptation in coastal areas.	OWOW
	SA27: Support climate readiness of coastal communities, including hazard mitigation, pre-disaster planning, preparedness, and recovery efforts.	OWOW
	SA28: Support preparation and response planning for impacts to coastal aquatic environments.	OWOW
Goal 11: Protect ocean environments by incorporating shifting environmental conditions and other emerging threats into EPA programs.	**SA29:** Consider climate change impacts on marine water quality in NWP ocean management authorities, policies, and programs.	OWOW
	SA30: Use available authorities and work with the Regional Ocean Organizations and other federal and state agencies through regional ocean groups and other networks so that offshore renewable energy production does not adversely affect the marine environment.	OWOW (Regions)
	SA31: Support the evaluation of sub-seabed sequestration of CO_2 and any proposals for ocean fertilization.	OWOW
	SA32: Participate in interagency development and implementation of federal strategies through the NOC and the NOC Strategic Action Plans.	OWOW
Water Quality: Our Nation's surface water, drinking water, and ground water quality are protected, and the risks of climate change to human health and the environment are diminished, through a variety of adaptation and mitigation strategies.		
Goal 12: Protect waters of the United States and promote management of sustainable surface water resources.	**SA33:** Encourage states and communities to incorporate climate change considerations into their water quality planning.	OWOW
	SA34: Encourage green infrastructure and low-impact development to protect water quality and make watersheds more resilient.	OWM (OWOW)
	SA35: Promote consideration of climate change impacts by National Pollutant Discharge Elimination System permitting authorities.	OWM
	SA36: Encourage water quality authorities to consider climate change impacts when developing wasteload and load allocations in TMDLs where appropriate.	OWOW

Goals and Strategic Actions		Lead Office (& Partners)
	SA37: Identify and protect designated uses that are at risk from climate change impacts.	OST (OWM)
	SA38: Clarify how to re-evaluate aquatic life water quality criteria on more regular intervals; and develop information to assist states and tribes who are developing criteria that incorporate climate change considerations for hydrologic condition.	OST
Goal 13: As the nation makes decisions to reduce greenhouse gases and develop alternative sources of energy and fuel, work to protect water resources from unintended adverse consequences.	**SA39:** Continue to provide perspective on the water resource implications of new energy technologies.	OWM (OGWDW)
	SA40: Provide assistance to states and permittees to assure that geologic sequestration of CO_2 is responsibly managed.	OGWDW (OWOW)
	SA41: Continue to work with States to help them identify polluted waters, including those affected by biofuels production, and help them develop and implement Total Maximum Daily Loads (TMDLs) for those waters.	OGWDW (OWOW, OWM)
	SA42: Provide informational materials for stakeholders to encourage the consideration of alternative sources of energy and fuels that are water efficient and maintain water quality.	OWM (OW IO)
	SA43: As climate change affects the operation or placement of reservoirs, work with other federal agencies and EPA programs to understand the combined effects of climate change and hydropower on flows, water temperature, and water quality.	OWM
Goal 14: Collaborate to make hydrological and climate data and projections available.	**SA44:** Monitor climate change impacts to surface waters and ground water.	OWOW (OGWDW)
	SA45: Collaborate with other federal agencies to develop new methods for use of updated precipitation, storm frequency, and observational streamflow data, as well as methods for evaluating projected changes in low flow conditions.	OW IO
	SA46: Enhance flow estimation using National Hydrography Dataset Plus (NHDPlus).	OWOW
Working With Tribes: Tribes are able to preserve, adapt, and maintain the viability of their culture, traditions, natural resources, and economies in the face of a changing climate.		
Goal 15: Incorporate climate change considerations in the implementation of core programs, and collaborate with other	**SA47:** Through formal consultation and other mechanisms, incorporate climate change as a key consideration in the revised NWP Tribal Strategy and subsequent implementation of CWA, SDWA, and other core programs.	OW IO
	SA48: Incorporate adaptation into tribal funding mechanisms, and collaborate with other EPA and federal funding programs to support sustainability and adaptation in tribal communities.	OW IO

(Continued)

Goals and Strategic Actions		Lead Office (& Partners)
EPA Of fices and federal agencies to work with tribes on climate change issues on a multi-media basis.		
Goal 16: Tribes have access to information on climate change for decision making.	**SA49:** Collaborate to explore and develop climate change science, information, and tools for tribes, and incorporate local knowledge.	OW IO
	SA50: Collaborate to develop communication materials relevant for tribal uses and tribal audiences.	OW IO
Cross-Cutting Program Support		
Goal 17: Communicate, Collaborate, and Train.	**SA51:** Continue building the communication, collaboration, and training mechanisms needed to effectively increase adaptive capacity at the federal, tribal, state, and local levels.	OW IO
Goal 18: Track Progress and Measure Outcomes	**SA52:** Adopt a phased approach to track programmatic progress towards Strategic Actions; achieve commitments reflected in the Agency Strategic Plan; work with the EPA Work Group to develop outcome measures.	OW IO
Goal 19: Identify Climate Change and Water Research Needs	**SA53:** Work with ORD, other water science agencies, and the water research community to further define needs and develop research opportunities to deliver the information needed to support implementation of this *2012 Strategy*, including providing the decision support tools needed by water resource managers.	OST (OW IO)

Appendix C: List of Abbreviations

ACWI	Advisory Committee on Water Information	**LCCs**	Landscape Conservation Cooperatives
ASR	Aquifer Storage & Recovery	**LID**	Low Impact Development
AWWA	American Water Works Association	**LIDAR**	Light Detection And Ranging
BLM	Bureau of Land Management	**MPRSA**	Marine Protection, Research and Sanctuaries Act
BOEMR	Bureau of Ocean Energy Management, Regulation& Energy	**MWDs**	Municipal Water Districts
CCAWWG	Climate Change Adaptation Work Group	**NAP**	National Adaptation Plan
CCL	Contaminant Candidate List	**NDWAC**	National Drinking Water Advisory Council
CEQ	White House Council on Environmental Quality	**NEP**	National Estuary Program
CMSP	Coastal Marine Spatial Planning	**NEPA**	National Environmental Policy Act
CRE	Climate Ready Estuaries	**NHDPlus**	National Hydrography Dataset Plus
CREAT	Climate Resilience Evaluation and Awareness Tool	**NMFS**	National Marine Fisheries Service
CRWU	Climate Ready Water Utilities	**NOAA**	National Oceanic and Atmospheric Administration
CT4CW	Coming Together for Clean Water	**NOC**	National Ocean Council
CWA	Clean Water Act	**NPDES**	National Pollutant Discharge Elimination System
DMR	Discharge Monitoring Report	**NPDWR**	National Primary Drinking Water Regulations
DOE	Department of Energy	**NPS**	National Park Service
DOT	Department of Transportation	**NRC**	National Research Council
DWSRF	Drinking Water State Revolving Fund	**NWCA**	National Wetland Condition Assessment
EAT	Energy Audit Tool	**NWI**	National Wetland Inventory
FEMA	Federal Emergency Management Agency	**NWP**	National Water Program
GAP	General Assistance Program	**OAR**	Office of Air and Radiation
GHG	Greenhouse gas	**OCSPP**	Office of Chemical Safety and Pollution Prevention
GI	Green Infrastructure	**OGC**	Office of General Counsel
HAB	Harmful algal bloom	**OGWDW**	Office of Ground Water and Drinking Water (OW)
HUC	Hydrologic Unit Code	**OITA**	Office of International and Tribal Affairs
HUD	Housing and Urban Development	**OM&R**	Operations, maintenance and replacement

HWQS	Hydrologic and Water Quality System	**OP**	Office of Policy
ICCATF	Interagency Climate Change Adaptation Task Force	**ORD**	Office of Research and Development
IPCC	Intergovernmental Panel on Climate Change	**OST**	Office of Science and Technology(OW)
IWRM	Integrated Water Resources Management	**TMDLs**	Total Maximum Daily Loads
OSTP	White House Office of Science & Technology Policy	**USFWS**	U.S. Fish and Wildlife Service
OSWER	Office of Solid Waste and Emergency Response	**UIC**	Underground Injection Control
OUST	Office of Underground Storage Tanks (OSWER)	**USACE**	U.S. Army Corps of Engineers
OW	Office of Water	**USBR**	U.S. Bureau of Reclamation
OWM	Office of Wastewater Management(OW)	**USDA**	U.S. Department of Agriculture
OWOW	Office of Wetlands, Oceans and Watersheds (OW)	**USDW**	Underground source of drinking water
RISAs	Regional Integrated Sciences and Assessments	**USFS**	U.S. Forest Service
SAP	Synthesis and Assessment Product	**USGCRP**	U.S. Global Change Research Program
SDWA	Safe Drinking Water Act	**USGS**	U.S. Geological Survey
SWAQ	Subcommittee on Water Availability and Quality	**WQS**	Water quality standards
		WUE	Water Use Efficiency

Appendix D: References

[1] BOR, (2010). Memorandum of Understanding for Hydropower, Among the Department of Interior, Department of Energy, and the Army, March 24, 2010. Available at: http://www.usbr.gov/power/SignedHydropower MOU.pdf.

[2] BOR, (2011). SECURE Water Act Section 9503(c) Reclamation Climate Change and Water 2011, April 2011. Available at: http://www.usbr.gov/climate/SECURE/.

[3] Brekke, (2009). *Climate change and water resources management—A federal perspective: U.S. Geological Survey Circular 1331*, Brekke, L.D., Kiang, J.E., Olsen, J.R., Pulwarty, R.S., Raff, D.A., Turnipseed, D.P., Webb, R.S., and White, K.D., 2009. Available at: http://pubs.usgs.gov/circ/1331/.

[4] CA, (2011a). IWRM Handbook, California Department of Water Resources. Available at: www.water.

[5] CA, (2011b). *Embedded Energy in Water Pilot Programs Impact Evaluation Final Report*, ECONorthwest, March 9, 2011. Prepared for the California Public Utilities Commission Energy Division. Available at: http://www.energydataweb.com/cpucFiles/33/FinalEmbeddedEnergy PilotEMVReport_1.pdf.

[6] CBPO, (2010). Chesapeake Bay Executive Order Strategy, May 12, 2010. Available at: http://executiveorder.chesapeakebay.net/post/New-Federal-Strategy-for-Chesapeake-LaunchesMajor-Initiatives-and-Holds-Government-Accountable-for-Progress.aspx.

[7] CCAWWG, (2011). *Addressing Climate Change in Long-Term Water Resources Planning and Management: User Needs for Improving Tools and Information*. Available at: http://www.usbr.gov/research/ climate.

[8] CEQ, (2009). Executive Order 13514 *Federal Leadership in Environmental, Energy, and Economic Performance*," Section 16. October 5, 2009. Available at: http://www.whitehouse.gov/assets.

[9] CEQ, (2010). Interagency Climate Change Adaptation Task Force, report to the President, October 4, 2010, and Water Workgroup Report, October 5, 2010. Available at: http://www.whitehouse.gov/ administration/eop/ceq/initiatives/adaptation.

[10] CEQ, (2011a). *National Action Plan: Priorities for Managing Freshwater Resources in a Changing Climate*, Report of the Freshwater Workgroup, Interagency Climate Change Adaptation Task Force, October 28, 2011. Available at: http://www.whitehouse.gov/sites/ default/files/ microsites/ceq/2011_national_action_plan.pdf.

[11] CEQ, (2011b). *Instructions for Implementing Climate Change Adaptation Planning In Accordance With Executive Order 13514, Federal Agency Climate Change Adaptation Planning*

[12] *Implementing Instructions*, March 4, (2011). Available at: http://www.whitehouse.gov/sites/default/files/microsites/ceq/adaptation.

[13] CSO, (2011). Coastal States Organization web site, accessed January, 2011. Available at: http://www.coastalstates.org/.

[14] DOE, (2011). U.S. Department of Energy, Energy Efficiency and Renewable Energy. Available at: http://www.eere.energy.gov/.

[15] DOE, (2012). Secure Water Act of 2009 (Public Law 111-11), Section 9505 Report to Congress on the effects of climate change on federal hydropower systems. *Forthcoming.* Available at: http://nhaap.ornl.gov/ content/climate-change-impacts.

[16] Dore, M. & Burton, I. (2001). The Costs of Adaptation to Climate Change in Canada: A Stratified Estimate by Sectors and Regions—

Social Infrastructure. Climate Change Laboratory, Brock University, St. Catharines, Ontario. 117 pp.

[17] EPA, (2011). Environmental Protection Agency Climate Change website. Accessed June 13, 2011.Available at: http://www.epa.gov/climatechange/.

[18] EPA, (2004). *Guidelines for Water Reuse*. Environmental Protection Agency, EPA/625/R-04/108, September 2004. Available at: http://www.epa.gov/ord/NRMRL/pubs/625r04108/625r04108.pdf.

[19] EPA, (2007). Distribution System Inventory, Integrity and Water Quality. Environmental Protection Agency, January 2007. Available at: http://www.epa.gov/ogwdw/disinfection.

[20] EPA, (2008a). *National Water Program Strategy: Response to Climate Change*. Environmental Protection Agency, EPA 800-R-08-001, September 2008. Available at: http://water.

[21] EPA, (2008b). *Ensuring a Sustainable Future: An Energy management Guidebook for Wastewater and Water Utilities*. Environmental Protection Agency, January 2008. Available at: http:// water.

[22] EPA, (2009). Endangerment and Cause or Contribute Findings for Greenhouse Gases Under Section 202(a) of the Clean Air Act Final Rule, 74 Fed. Reg. 66496 (Dec. 15, 2009). Available at: http://epa.gov/climatechange/Downloads/endangerment/Federal_Register-EPA-HQ-OAR-2009-0171-Dec.15-09.pdf.

[23] EPA, (2009a). Biofuels Compendium. Environmental Protection Agency, July 21, 2009. Available at: http://www.epa.gov/oust/altfuels/bfcompend.htm.

[24] EPA, (2009b). National Pollutant Discharge Elimination System (NPDES). Environmental Protection Agency, March 12, 2009. Available at: http://cfpub.epa.gov/npdes/.

[25] EPA, (2009c). National Drinking Water Advisory Council Request for Climate Ready Water Utilities Working Group Nominations. Federal Register, July 8, 2009 (Volume 74, Number 129, pp. 32595 – 32596. Available at: http://www.epa.gov/fedrgstr/EPA-WATER/2009/July/Day-08/ w16006.htm.

[26] EPA, (2009d). U.S. EPA Proceedings: *First National Expert and Stakeholder Workshop on Water Infrastructure Sustainability and Adaptation to Climate Change*. Environmental Protection Agency. Available at: http://www.epa.gov/nrmrl/wswrd/wq/wrap/workshop.html.

[27] EPA, (2010a). FY 2011-2015 Strategic Plan. Environmental Protection Agency, September 30, 2010. Available at: http://www.epa.gov/planandbudget/strategicplan.html.

[28] EPA, (2010b). EPA Guidelines for Preparing Economic Analyses, Dec. 2010. Available at: http://yosemite.epa.gov/ee/epa/eed.nsf/webpages.

[29] EPA, (2010c). *Clean Water and Drinking Water Infrastructure Sustainability Policy*, Oct. 4, 2010. Available at: http://water.

[30] EPA, (2010d). *Control and Mitigation of Drinking Water Losses in Distribution Systems*. Environmental Protection Agency, EPA 816-R-10-019, November 2010. Available at: http://www. epa.gov/region9/waterinfrastructure/waterenergy.html.

[31] EPA, (2010e). Underground Injection Control Program. Environmental Protection Agency, December 13, 2010. Available at: http://water.epa.gov/type/groundwater/uic/.

[32] EPA, (2010f). *Compatibility of Underground Storage Tank Systems with Biofuel Blends*. Federal Register, November 17, 2010 (Volume 75, Number 221, pp. 70241 – 70246. Available at: http://www.gpo.gov/fdsys/pkg/FR-2010-11-17/pdf/2010-28968.pdf.

[33] EPA, (2010g). Guidelines on Water Efficiency Measures for Water Supply Projects in the Southeast, USEPA Region 4, June 21, 2010. Available at: http://www.epa.gov/region4/water.

[34] EPA, (2011a). Policy Memo from Lisa Jackson establishing the EPA Climate WG. http://www. epa.gov/climatechange/effects/downloads/adaptation.

[35] EPA, (2011b). *Coming Together for Clean Water: EPA's Strategy To Protect America's Waters*. April 2011. Available at: https://blog.epa.gov/waterforum/wp-content/uploads/2011/04/ComingTogether-for-Clean-Water-FINAL.pdf.

[36] EPA, (2011c). Climate Ready Water Utilities Toolbox. Available at: http://www.epa.gov/safewater/water.

[37] EPA, (2011d). Sustainable Infrastructure: Energy Efficiency & Renewable Energy Opportunities, Environmental Protection Agency. Accessed June 12, 2011. Available at: http://water.

[38] EPA, (2011e). WaterSense Program Accomplishments, Environmental Protection Agency. Available at: http://www.epa.gov/WaterSense/about_us/program_accomplishments.html.

[39] EPA, (2011f). WaterSense, Environmental Protection Agency. Available at: http://www.epa.gov/ WaterSense/.

[40] EPA, (2011g). Healthy Watersheds. Environmental Protection Agency, April 5, 2011. Available at: http://water.epa.gov/polwaste/nps/watershed/index.cfm.

[41] EPA, (2011h). *Rolling Easements Primer*. Environmental Protection Agency, June, 2011. Available at: www.epa.gov/cre/downloads/rollingeasementsprimer.pdf.

[42] EPA, (2011i). Memorandum of Agreement. DHS (FEMA and EPA. Environmental Protection Agency, May 12, 2010. Available at : http://www.epa.gov/dced/pdf/2011_0114_fema-epa-moa. pdf.

[43] EPA, (2011j). *Managing Wet Weather With Green Infrastructure*. Environmental Protection Agency, January 4, 2011. Available at: http://cfpub.epa.gov/npdes/home.cfm?program_id=298.

[44] EPA, (2011k). Low Impact Development. Environmental Protection Agency, March 18, 2011. Available at: http://www.epa.gov/owow/NPS/lid/.

[45] EPA, (2011l). Impaired Waters and Total Maximum Daily Loads. Environmental Protection Agency, March 21, 2011. Available at: http://water.

[46] EPA, (2011m). Renewable Fuels: Regulations and Standards. Environmental Protection Agency, March 23, 2011. Available at: http://www.epa.gov/oms/fuels/renewablefuels/regulations.htm.

[47] EPA, (2011n). National Aquatic Resource Surveys. Environmental Protection Agency, April 8, 2011. Available at: http://water.epa.gov/type/watersheds/monitoring/nationalsurveys.cfm.

[48] EPA, (2012a). EPA Climate Change Website, Impacts and Adaptation page. Available at: http://epa.gov/climatechange/impacts-adaptation/index.html.

[49] EPA, (2012b). Energy Use Assessment Tool. Available at: http://water.epa.gov/infrastructure/sustain/energy_use.cfm.

[50] EPA, (2012c). EPA Green Infrastructure Strategy web site can be found at: http://water.epa.gov/infrastructure/greeninfrastructure/index.cfm.

[51] EPA, (2012d). Integrated Municipal Stormwater and Wastewater Plans. EPA policy memos. Available at: http://cfpub.epa.gov/npdes/integratedplans.cfm.

[52] EPA-OAR, (2011). Energy Efficiency & Renewable Energy Opportunities. Website, accessed June 18, 2011. Available at: http://water.epa.gov/infrastructure/sustain/energyefficiency.cfm.

[53] EPA-ORD. Science Inventory. Searchable database. Available at: http://cfpub.epa.gov/si/.

[54] EPA-R9, (2011). Sustainable Water Infrastructure. Environmental Protection Agency, Region 9, December 31, 2010. Available at: http://www.epa.gov/region9/waterinfrastructure/index.html.

[55] EPRI, (2002). *Water & Sustainability (Volume 4): U.S. Electricity Consumption for Water Supply & Treatment – the next half century.* Electric Power Research Institute (EPRI), 2002.

[56] Frederick, (2000). Frederick, K. and G. Schwarz. 2000. Socioeconomic Impacts of Climate Variability and Change on U.S. Water Resources. Washington, DC: Resources for the Future.

[57] FWP, (2011). Fish, Wildlife and Plants Climate Adaptation Workgroup. Available at: http://www. wildlifeadaptationstrategy.gov/index.php.

[58] Hayhoe, K., Wake, C. P., Huntington, T. G., Luo, L., Schwartz, M., Sheffield, J., Wood, E., Anderson, B., Bradbury, J., DeGaetano, A., Troy, T. & Wolfe, D. (2007). Climate Dynamics. *Past and Future Changes in Climate and Hydrological Indicators in the U.S. Northeast.* Available at: http://www.northeastclimateimpacts.org/pdf/tech/hayhoe_et_al_climate_dynamics_2006.pdf.

[59] ICWP, (2012). Interstate Council on Water Policy. Website: http://www.icwp.org.

[60] Jackson, (2006). Jackson, S. Vegetation, environment, and time: the origination and termination of ecosystems. *Journal of Vegetation Science*, *17*, 549-557. Available at: http://www. bioone.org/doi/abs/10.1658/1100-9233(2006)17%5B549:VEATTO%5D2.0.CO%3B2.

[61] Kirshen, Kirshen, R., Ruth, M. & Anderson, W. (2006). "*Climate's Long term Impacts on Urban Infrastructures and Services: The Case of Metro Boston.*" In: Regional Climate Change and Variability: Impacts and Responses, Eds. M. Ruth, K. Donaghy, and P. Kirshen.

[62] Kirshen, Kirshen, P., Watson, C., Douglas, E., Gontz, A., Lee, J. & Tian, Y. (2008). Mitigation and Adaptation Strategies for Global Change, *Coastal Flooding in the Northeastern United States Due to Climate Change*. 2008. Available at: http://www.northeastclimateimpacts.org/pdf/miti/kirshen_et_al.pdf.

[63] MEA, (2005). Millennium Ecosystem Assessment. Ecosystems and Human Well-being: Synthesis. Washington, DC: Island Press, 2005.

[64] MWDSC, (2008). Integrated Resources Plan (IRP). Metropolitan Water District of Southern California, October 2008. Available at: http://www.mwdh2o.com/mwdh2o/pages/yourwater/irp/More_Water_Resources.html.

[65] Neumann, Neumann, J., Hudgens, D., Herter, J., Martinich, J. (2010). *The economics of adaptation along developed coastlines*. Published Online: Dec 08 2010. Available at: http://wires.

[66] NDWAC, (2010). *Climate Ready Water Utilities Working Group, Final Report to the National Drinking Water Advisory Council*. Dec. 9, 2010. Available at: http://water.

[67] NOC, (2012). The National Ocean Council. Available at: http://www.whitehouse.gov/administration/eop/oceans.

[68] NRC, (2009). *Ecological Impacts of Climate Change*. National Academy of Sciences. Available at: http://dels-old.nas.edu/dels/rpt_briefs/ecological_impacts.pdf.

[69] NRC, (2010a-d). *America's Climate Choices*. Four Volumes: a) *Limiting the Magnitude of Future Climate Change*; b) *Adapting to the Impacts of Climate Change*; c) *Advancing the Science of Climate Change*; d) *Informing Effective Decisions and Actions Related to Climate Change*. National Academies of Sciences, 2010. Available at: http://Americas climatechoices.org/.

[70] NRC, (2010e). *Climate Stabilization Targets: Emissions, Concentrations, and Impacts Over Decades to Millennia*. http://books.nap.edu/catalog.php?record_id=12877.

[71] NRC, (2010f). *Ocean Acidification: A National Strategy to Meet the Challenges of a Changing Ocean*. National Academies Press, ISBN: 978-0-309-15359-1. Available at: http://www.nap.edu/ catalog.php?record_id=12904.

[72] NRC, (2011a). *Warming World: Impacts by Degree*. Based on the National Research Council report, *Climate Stabilization Targets: Emissions, Concentrations, and Impacts over Decades to Millennia* (2011). Available at: http://dels.nas.edu/materials/booklets/warming-world.

[73] OCWD, (2008). Orange County Water District Programs and Projects Overview. Accessed June 13, 2011. Available at: http://www.ocwd.com/Programs---Projects/ca-17.aspx.

[74] Opperman, (2010). Opperman, Jeffrey, Luster, Ryan, et.al. Ecologically Functional Floodplains: Connectivity, Flow Regime, and Scale. *Journal of the American Water Resources Association*, *46*(2), 211-226, 2010.

[75] TNC, Smith, M. P., Schiff, R., Olivero, A. & MacBroom, J. (2008). The Active River Area: A Conservation Framework for Protecting Rivers and Streams. Boston: The Nature Conservancy, 2008. Available at: http://www.floods.org/PDF/ASFPM_TNC_Active_River_%20Area.pdf.

[76] UK DEFRA, (2010). Adapting to Climate Change: Guidance notes for NI188, March 2010. Available at: http://www.defra.gov.uk/ environment/climate/sectors/local-authorities/; and http://archive. defra. gov.uk/corporate/about/with/localgov/indicators/documents/ni188-guidance.

[77] USACE, (2010a). *Building Strong Collaborative Relationships for a Sustainable Water Resources Future: National Report Responding to National Water Resources Challenges*, August, 2010. U.S. Army Corps of Engineers. Available at: http://www.building-collaboration.

[78] USACE, (2010b). Federal Support Toolbox for IWRM. Available at: http://www.building-collaboration.

[79] USBR, (2005). *Water 2025: Preventing Crises and Conflict in the West*. U.S. Bureau of Reclamation, Washington, D.C., 32 pp. Available at: http://permanent.access.

[80] USBR, (2011). WaterSMART: Sustain and Manage America's Resources for Tomorrow. U.S. Department of the Interior, Bureau of Reclamation, March 22, 2011. Available at: http://www.usbr.gov/ WaterSMART/.

[81] USGCRP, (2008). Synthesis and Assessment Product 4.4: *Preliminary review of adaptation options for climate-sensitive ecosystems and resources*. U.S. Global Change Research Program. Available at: http:// www.climatescience.gov/Library/sap/sap4-4/final-report/.

[82] USGCRP, (2009a). *Global Climate Change Impacts in the United States*, Thomas R. Karl, Jerry M. Melillo, and Thomas C. Peterson, (eds.). Cambridge University Press, 2009. Available at: http://www. globalchange.gov/publications/reports/scientific-assessments/us-impacts.

[83] USGCRP, (2009b). Synthesis and Assessment Product 4.1: *Coastal Sensitivity to Sea-Level Rise: A Focus on the Mid-Atlantic Region*. U.S. Global Change Research Program. Available at: http://www. globalchange.gov/publications/reports/scientific-assessments/saps/302.

[84] USGCRP, (2011). National Climate Assessment Workshop, Draft Report: *Valuation Techniques and Metrics for Climate Change Impacts, Adaptation, and Mitigation Options*. Available at: http://www.nesdis. noaa.gov/NCADAC/pdf/15b.pdf.

[85] USGCRP, (2012). U.S. Global Change Research Program, National Climate Assessment. Information available at: http://www.globalchange. gov/index.php?option=com_content&view=article&id=417&Itemid= 401.

[86] USGS, (2011). National Hydrography Dataset Plus website. Accessed June 13, 2011. Available at: http://www.horizon-systems.com/nhdplus/.

[87] WERF, (2010). *Sustainable Treatment: Best Practices from the Strass im Zillertal Wastewater Treatment Plant*, March, 2010. Water Environment Research Foundation, Stock No. OWSO4R07b. Available at: http://www.werf.org/AM/Template.cfm?Section=Search&Template=/CustomSource/Research/PublicationProfile.cfm&id=OWSO4R07b.

[88] WRF, (2011). The Future of Research on Climate Change. Impacts on Water. A Workshop Focusing on Adaptation Strategies and Information Needs. Available at: http://www.waterrf.org/projectsreports/publicreportlibrary/4340.pdf.

[89] World Resources Institute, (2005). Millennium Ecosystem Assessment (MEA), 2005. Available at: http://www.maweb.org/documents/document.358.aspx.pdf.

Appendix E: Acknowledgements

This *2012 Strategy* is the collective product of many people. Special thanks go to the following people who contributed significantly to the development of this document.

National Water Program Workgroup Members

EPA Office of the Assistant Administrator for Water

David Bylsma, Joel Corona, Elana Goldstein, Patrick Maloney, Karen Metchis, John Powers, Michael Shapiro, Felicia Wright

Office of Groundwater and Drinking Water

Curt Baranowski, Rachel Herbert, Keara Moore, Mike Muse, David Travers, John Whitler

Office of Wetlands, Oceans and Watersheds

Michael Craghan, Paul Cough, Holly Elwell, Laura Gabanski, Kathleen Kutschenreuter, Bernice Smith

Office of Wastewater Management

Veronica Blette, Caitlin Gregg, Randy Hill, Sarita Hoyt

Office of Science and Technology

Rachael Novak

EPA Regional Offices

Region 1: Mel Cote, Ken Moraff, Stephen Perkins
Region 2: Douglas Pabst, Patricia Pechko
Region 3: Joe Piotrowski
Region 4: Bob Howard, Linda Rimer
Region 5: Kate Balasa, Tinka Hyde
Region 6: James R. Brown, Barbara Keeler
Region 7: Karen Flournoy, Morris Holmes, Mary Mindrup
Region 8: Jim Berkley, Carol Russell
Region 9: Michael Mann, Suzanne Marr, Cheryl McGovern, Karen Schwinn
Region 10: Paula VanHaagen, Sharon Wilson
Great Lakes Program: John Haugland
Gulf of Mexico Program: John Bowie

Other EPA Contributors

EPA Office of Air and Radiation

- Rona Birnbaum
- Jeremy Martinich
- William Perkins

EPA Office of Policy

- Catherine Allen
- Leah Cohen
- Gerald Filbin
- Joel Scheraga

EPA Office of Research and Development

- James Goodrich
- Anne Grambsch
- Tom Johnson
- Andy Miller
- Jennifer Orme-Zavaleta

- Suzanne VanDrunick
- Chris Weaver
- Jeff Yang

Special Acknowledgments for Partners and Stakeholders

The NWP owes a debt of gratitude to the many people who have engaged in the dialogue on climate change and water resources. While there are too many to name them all, we especially want to express our appreciation to the following organizations and individuals whose ideas have informed our thinking in the process of developing this *2012 Strategy*, and who have created a collaborative problem-solving environment that serves the nation well in tackling this complex issue.

State-Tribal Climate Change Council

Association of State Drinking Water Administrators (ASDWA)

- Denise Clifford, Washington Department of Health
- Jessica Godreau, North Carolina Department of Environment and Natural Resources
- Elston Johnson, Texas Commission on Environmental Quality
- Saeid Kasraei, Maryland Department of the Environment
- Dierdre Mason, ASDWA
- Fred Sickel, New Jersey Department of Environmental Protection
- Jim Taft, ASDWA

Association Clean Water Administrators (ACWA, formerly ASIWPCA)

- Dave Akers, Colorado Department of Public Health and Environment
- Arthur Baggett, California State Water Resources Control Board
- Carol Collier, Delaware River Basin Commission
- Alexandra Dunn, ACWA
- Susan Kirsch, ACWA
- Janet Llewellyn, Florida Department of Environmental Protection
- Sara Vinson, ACWA
- Rebecca Weidman, New England Interstate Water Pollution Control Commission
- Marcia Willhite, Illinois Environmental Protection Agency

Association of State Wetland Managers (ASWM)

- Anna Buckley, Oregon Department of State Lands
- Jeanne Christie, ASWM
- Denise Clearwater, Maryland Department of the Environment
- Doug Fry, Florida Department of Environmental Protection
- Ted LaGrange, Nebraska Game and Parks Commission
- Alan Quakenbush, Vermont Department of Environmental Conservation

Ground Water Protection Council (GWPC)

- Mark Fesmire, New Mexico Energy, Minerals, and Natural Resources Department
- Hal Fitch, Michigan Department of Environmental Quality
- Joe Lee, Pennsylvania Department of Environmental Protection
- Mike Paque, GWPC
- Sarah Pillsbury, New Hampshire Department of Environmental Services
- Dave Terry, Massachusetts Department of Environmental Protection

National Tribal Water Council (NTWC)

- Michael Bolt, Eastern Band of Cherokee Nation
- Alex Cabillo, Arizona Hualapai Tribe
- Daniel Chythlook, Aleknagik Traditional Council

- Steve Crawford, Passamaquoddy Tribe at Pleasant Point
- Dave Fuller, Port Gamble S'Klallam Tribe
- Nancy John, Cherokee Nation

EPA Appointed Members

- Steve Etsitty, Navajo Nation EPA
- Denise Jensen, Winnebago Tribe of Nebraska
- Jeanine Jones, Western Governors Association/California Department of Water Resources
- Lynn Kent, Alaska Department of Environmental Conservation

- Chief Oren Lyons, Onondaga Nation, Haudenosaunee Confederacy
- Naomi Tillison, Bad River Department of Natural Resources
- Jason Walker, Northwestern Band of the Shoshone Nation

Association Partners

The Water Environment Research Foundation (WERF) and the Water Research Foundation (WaterRF), which conduct research and development activities related to wastewater treatment utilities and drinking water utilities, respectively.

The American Water Works Association (AWWA) and the Association of Metropolitan Water Agencies (AMWA), which provided valuable input and support throughout this process.

The Water Utility Climate Alliance (WUCA) members, who have provided leadership on climate change.

Climate Ready Water Utilities Working Group of the National Drinking Water Advisory Committee for their leadership and commitment to this issue.

End Notes

[1] The term "National Water Program" refers to the Office of Water (OW) plus the water programs in the 10 EPA Regions, and recognizes that many of our programs are implemented by state and tribal water authorities.

[2] Since 2009, Michael Shapiro, EPA Deputy Assistant Administrator for Water, has served as co-chair of the Task Force's Water Workgroup along with Matthew Larsen, U.S. Geological Survey (USGS) Associate Director for Climate and Land Use Change, and Jeffrey Peterson, White House Council on Environmental Quality Deputy Associate Director for Water Policy.

[3] Printed copies of the National Action Plan are available by sending an email to the following address: adaptation@ceq.eop.gov, stating addressee, mailing address, and the number of copies desired (limit of three).

[4] This list is intended to be illustrative of recent published research. EPA is not endorsing any specific estimate.

[5] Section 604(b) of the CWA establishes a grant program to fund state, local, and interstate water quality planning efforts under CWA sections 205(j) and 303(e). This provision requires states to reserve 1% of their Clean Water State Revolving Fund allotment, or $100,000, whichever is greater, for planning. Under section 205(j), many states pass through at least 4 0% of these funds to local or interstate planning organizations.

[6] OGWDW=Office of Groundwater and Drinking Water; OWM=Office of Wastewater Management; OWOW=Office of Wetlands, Oceans and Watersheds; OST=Office of Science and Technology; OW IO=Office of Water Immediate Office

In: National Water Program Strategy … ISBN: 978-1-62618-124-3
Editor: Jun Sonoda

Chapter 2

FREQUENTLY ASKED QUESTIONS, NATIONAL WATER PROGRAM STRATEGY: RESPONSE TO CLIMATE CHANGE*

United States Environmental Protection Agency

Q1. Why Are You Revising the Previous EPA National Water Program Climate Strategy?

A: When EPA developed the first National Water Program Climate Strategy in 2007 (published in 2008), the Agency was starting to have conversations about the impacts of climate change on the EPA's water programs. EPA identified a range of actions to help begin the process of building the Agency's knowledge on climate change adaptation. Much has been learned since then, and EPA decided it was time to lay out a longer-term strategy to guide program planning that goes beyond the next two or three years.

In addition, there is now much more activity underway throughout the U.S. in local, tribal, state and federal areas.

This presents EPA with the opportunity to build stronger collaborations needed to achieve its goals. This Strategy attempts to elaborate on potential collaborations.

* This is an edited, reformatted and augmented version of the United States Environmental Protection Agency publication, dated December 2012.

Q2. What Is Your Main Message?

A: Climate change alters the hydrological background in which ecosystems and society function. To remain effective, EPA and its partners and stakeholders need to understand the changes underway in order to account for not only changes in average conditions but also changes in extreme conditions. However, while there is strong evidence that changes are underway, there is a range of uncertainty about the exact nature and extent of changes at the local scale where most decisions are made. Therefore, water managers need to develop flexible adaptive management strategies while we collectively continue to build the tools needed to project changes, understand vulnerability, and respond appropriately.

The pressures and stresses on the water environment come from many directions – from land use, population growth, municipal uses, industrial activity, agriculture, energy generation – all competing for limited supplies, all contributing to water quality impairment, and all struggling with how to account for the hydrological shifts associated with climate change.

This 2012 Strategy presents a comprehensive look at EPA's entire National Water Program, and describes the long-term vision that will involve EPA working in conjunction with our federal, state, tribal and local partners to achieve. It also outlines the long-term goals reflecting EPA's role and programmatic mission. The 2012 Strategy also lays out strategic actions that are the mid-term building blocks that should be undertaken in the next 3 to 8 years in order to work towards achieving those goals.

Taken as a whole, the 2012 Strategy lays out the directional intentions for EPA's programs as we work to mainstream climate change considerations into our core programs and as we work collaboratively with a wide range of partners.

Q3. How Is This "National Water Program 2012 Strategy: Response to Climate Change" Different from the Other Federal Climate Change Strategies and Adaptation Plans That Have Been, or Will Soon Be, Released?

A: The Interagency Climate Change Adaptation Task Force (Interagency Task Force) was established by President Obama in 2009. Over 20 federal agencies have been working together to develop recommendations to adapt to climate change impacts both domestically and internationally and have developed three cross-cutting adaptation plans. EPA was involved in all three plans associated with the Interagency Task Force, which are:

1) Freshwater National Action Plan
2) Fish, Wildlife and Plants Adaptation Plan
3) Oceans and Coasts Implementation Plan of the National Ocean Policy

In addition, the Interagency Task Force is also requiring each Federal Agency to develop adaptation plans addressing their operations, missions and programs. Consequently, EPA as a whole is also developing such a plan, which will be publicly available in the future.

The *2012 EPA National Water Program Climate Change Strategy*, which updates and revises an initial strategy published in 2008, is an in-depth treatment of just one of EPA's program areas and is coordinated with EPA's forthcoming Agency-wide climate adaptation plan. It also reflects EPA's commitments under the three national action plans mentioned above associated with the Interagency Task Force.

Q4. How Does the Strategy Address Stopping Global Warming?

A: The *2012 Strategy* describes opportunities to reduce greenhouse gases in the water sector, which accounts for 3% to 14% of the nation's energy use (depending on what you count). The strategy describes how EPA's National Water Program intends to promote opportunities to increase energy efficiency at wastewater and drinking water treatment plants, promote technologies to recover energy from waste streams, and reduce water consumption which then reduces energy needed to pump, treat, distribute and use water.

Q5. Does This Strategy Impose Any Regulations or Policies?

A: No. This document provides a bird's eye view of the EPA National Water Program and how it needs to evolve in order to be a 'climate resilient' water program. It provides directional intention, but it does not establish any binding commitments, regulations or policies. However, it does describe the challenges the program will face and how those programs should evolve over time. Particular policies and programs will be developed in the course of typical annual planning and action development processes.

Q6. How Is the Strategy Organized and What Are the Main Elements?

A: The principle theme of the 2012 Strategy is Integrated Water Resources Management (IWRM). IWRM is a framework to holistically address current water resource issues and emerging climate change complications. There are several definitions of this term, but for the purpose of this strategy, the NWP uses IWRM to describe opportunities for state, interstate, tribal and local officials (supported by Federal agencies) to voluntarily collaborate at watershed or aquifer scales to protect and preserve fresh water resources through mutually beneficial solutions. IWRM calls for inter-sector planning (e.g., between the energy sector and the water sector) to sustainably manage water resources.

The Strategy is organized around six key sections:

Infrastructure
Water Quality
Watersheds and Wetlands
Coastal and Ocean Waters
Working with Tribes
Geographic Regions

Q7. Why Do You Focus on "Working with Tribes" as a Separate Area?

A: Tribes have particular vulnerabilities due to climate change. Tribal cultures and economies are closely tied to the land they live on, and adaptation options may be limited. The *2012 Strategy* describes how EPA intends to work with Tribes to help them develop sustainable strategies for their communities.

Q8. Why Is the Regions Section of the Strategy Based on Geographic Regions Instead of EPA Regions?

A: EPA needs to work with Federal, state, tribal and local partners to achieve the visions and goals described in the *2012 Strategy*. Therefore, the strategy identifies the strategic issues shared by adjacent EPA Regions in order to leverage resources and build collaborations with other federal, state, tribal and local entities working at broader regional scales.

Q9. What Are Some of the Key Programs and Actions in the 2012 Strategy?

A:

- Climate Ready Estuaries (CRE)
- Climate Ready Water Utilities (CRWU)
- Green Infrastructure
- Healthy Watersheds Initiative
- Sustainable Infrastructure
- WaterSense
- Energy-Water Futures
- National Ocean Policy implementation
- Conduct research and develop information and tools to inform adaptation decisions

Q10. Why Are Non-EPA Programs Described in the 2012 Strategy?

A: There is a complex array of actors involved in managing water resources in the U.S. EPA's mission is to protect human health and the environment.

EPA's mandate is to protect the quality of surface water and aquatic ecosystems and protect drinking water, including underground sources of drinking water.

EPA is only involved in water quantity to the extent it is a factor in preserving aquatic ecosystems and water quality.

Yet, the inter-relationships between the water environment and the many uses of water by society are many, and the actors involved in managing water are diverse.

Therefore, to improve resilience of ecosystems and society, federal, state, tribal and local entities will need to work together.

Q11. How Do You Plan to Balance the Work Needed to Adapt to Climate Change with Existing Demands of Other Important Work of EPA, States and Tribes?

A: Over time water managers will learn how to mainstream climate change into their programs. This is a tough economic environment, and water managers are balancing many demands. Program managers will need to take the steps they can, including low-regret strategies that build flexibility for addressing multiple stressors including climate change. But it is also important to recognize that it is incumbent upon us to ensure that resources are not wasted and opportunities are not lost by failing to account for climate change impacts.

The actions described in the *2012 Strategy* will enhance our Nation's ability to sustainably manage water resources when facing a variety of stressors, including climate change.

Q12. How and When Will Progress Toward Goal Achievements Be Monitored?

The NWP intends to track our institutional process and progress in incorporating climate change considerations into EPA programs. An annual internal reporting process will assemble information for evaluating and publicly reporting progress. We have adopted a seven-phased system that recognizes the need for adaptive management that incorporates active learning and new information while managing risk. The seven phases are:

- Initiation
- Assessment
- Response Development
- Initial Implementation
- Robust Implementation
- Mainstreaming
- Monitoring and Adaptive Management

Highlights of progress will be publicly reported annually.

INDEX

C

P

Q

R

S

T

U

V

W